Cook Memorial Public Library

3 1122 01443 1840

P9-DCV-037

GARDEN FLORA

NOEL KINGSBURY

GARDEN

The Natural and Cultural History of the Plants in Your Garden

FLORA

COOK MEMORIAL LIBRARY
413 N. MILWAUKEE AVE.
LIBERTYVILLE, ILLINOIS 60048
JAN 4 - 2017

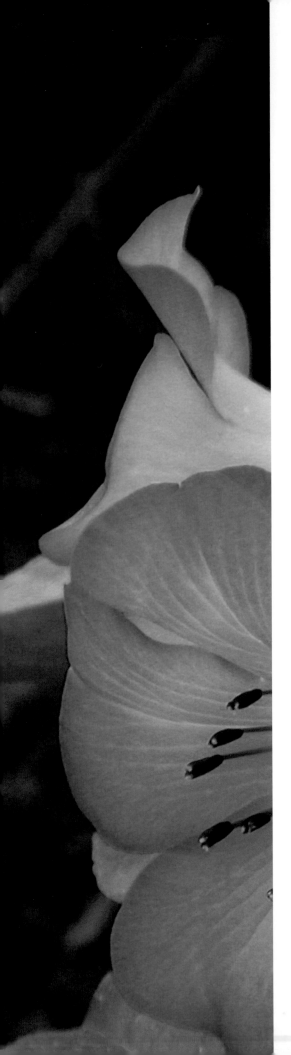

*In memory of Nicky Daw (1951–2016),
whose garden at Lower House, Cusop,
near Hay-on-Wye, Wales, created over many
years with her husband, Peter Daw,
gave inspiration to many.*

Copyright © 2016 by Noel Kingsbury. All rights reserved.

Thanks are offered to those who granted permission for use of materials but who are not named individually in the acknowledgments. While every reasonable effort has been made to contact copyright holders and secure permission for all materials reproduced in this work, we offer apologies for any instances in which this was not possible and for any inadvertent omissions.

Photo credits appear on page 353.

Published in 2016 by Timber Press, Inc.
The Haseltine Building
133 S.W. Second Avenue, Suite 450
Portland, Oregon 97204-3527
timberpress.com

Printed in China
Text design by Kristi Pfeffer
Jacket design by Patrick Barber

Library of Congress Cataloging-in-Publication Data

Names: Kingsbury, Noel, author.
Title: Garden flora: the natural and cultural history of the plants in your garden / Noel Kingsbury.
Other titles: Natural history of the plants in your garden
Description: Portland, Oregon: Timber Press, 2016. | Includes bibliographical references and index.
Identifiers: LCCN 2015047577 | ISBN 9781604695656 (hardcover)
Subjects: LCSH: Plants, Ornamental. | Botany.
Classification: LCC SB404.9 .K56 2016 | DDC 635.9—dc23
LC record available at http://lccn.loc.gov/2015047577

ISBN 13: 978-1-60469-565-6

A catalogue record for this book is also available from the British Library.

Introduction 8

Contents

Introduction

Read this first!!

Every garden plant comes from somewhere. It will have ancestral species growing on a mountainside or deep within a forest, or be part of some great remaining grassland. These ancestors may well have been put to some use by the local community, as food, medicine, or source of materials; they might have played a part in local mythology or spirituality. But at some point in history they were taken into cultivation and then truly become part of the human story. Some plants have been transformed by cultivation—selected, crossed, and bred to a level where their wild ancestors can hardly be recognised. Others have stayed remarkably similar to what might still be found in the world's diminishing wild spaces. But when were these plants introduced into cultivation? Where, when, and how did we end up with the range of cultivars we currently have down at the garden centre or nursery? This book is intended as an overview of the most widely grown genera of temperate zone garden plants, answering not just questions about origins and habitats but offering a broad outline of their history in cultivation as well.

Books on how to cultivate garden plants are plentiful, and many of these (as well as an increasing number of online sources) discuss the basic botanical characteristics of genera common in cultivation. These are two areas which are not dealt with here. What is much more difficult to access is information on the ecological aspects of garden plants and on their history in cultivation—areas that this volume focusses on. In addition, to help fill out the picture and provide some extra colour to our experience of garden plants, some reference is also made to the mythological and folklore aspects of garden plants, where relevant, and to their traditional and modern uses outside the garden.

The information available on the ecological and historical aspects of garden plants is very patchy: there is a lot on some genera, very little on others. This is partly a reflection of geography and development; it is easy, for example, to find material on the ecology of European plants, very difficult for Chinese. In some cases very detailed studies have been made of some genera, whereas others have been ignored. What I have tried to do is the classic "standing on the shoulders of giants"—reflecting upon and making available scholarship that is not available elsewhere, in one volume. In most cases, the genus is the unit of discussion; the exceptions are a number of headings that discuss a larger taxonomic unit or a cluster of genera (e.g., asters, gesneriads, heathers).

This introduction is aimed at talking the reader through some basic concepts, using keywords, which are here emphasised in bold. These keywords will then be used throughout the book as a shorthand. In many cases these concepts can only be understood in a wider context; consequently, the content here will act as a very basic primer in plant ecology and in garden plant history. Certain key personalities (plant hunters, botanists, etc.) who are repeatedly referenced, often by surname only, will also be indicated here in bold. Readers should be able to refer back to this introduction to find explanations of keywords.

The Genus

The first part of each entry in this book is designed to give the reader a broad overview of the genus: the origin of its name, its size (in terms of species) and native range, and the form the plants take. For gardeners who have got beyond the basics, the name of a plant genus is a kind of keystone for our understanding of plants, of making some sort of sense of them in our minds and for communicating with others about them. With complex genera, of major garden importance, there follows a breakdown of the often bewildering array of subcategories, which gardeners and botanists have used to try to bring some order to what often seems like chaos.

Wild plants exist as species; each has a binomial scientific name (genus + species), initially developed by the Swedish botanist Carl Linnaeus (1707–1778). Many garden plants, however, are cultivars, not natural species. **Cultivars** are often selections from a single species, of which every individual is genetically identical; many cultivars are **hybrids**—the result of a cross between two or more distinct lineages (which could be species, cultivars, or other defined and distinct groupings below species level). Taxonomic units—species, subspecies, cultivar—are collectively referred to as **taxa** (singular, **taxon**).

Ok? With this little bit of knowledge, much of the apparent complexity of plant naming falls into place. We could go on: the family is another very useful categorisation concept, and this too is stated for each plant entry. "Very useful categorisation concept"—actually I'm having second thoughts about this, as several times here I find myself referring to recent changes in families, a result of the advanced (and mind-numbingly complex) field of phylogeny (the study of evolutionary relationships), which is using DNA-based (and other) data to do a lot of reclassifying. Many familiar families are being reorganised in a way that goes against many of our deepest intuitions.

Plant growth habit

Trees, shrubs, climbers, annuals, biennials—all are well-understood words, although we will be discussing some qualifications. "Perennial" is a bit ambiguous, as technically it covers any plant form with a long lifespan; here we follow the popular parlance of using it for longer-lifespan herbaceous plants. **Herbaceous** refers to the habit of dying back every year in the winter, or the dry season, as opposed to woody plants, which accumulate growth from year to year. **Subshrubs** is not a botanical term but a very useful one for gardeners, as we seem to be drawn to these low-growing, semi-woody shrubs, with their pleasing rounded shapes and (nearly always) evergreen foliage—lavender is perhaps the best example of one. **Lianas** are large climbers with woody stems—effectively, trees that cannot stand up straight. They are what Tarzan swings on.

Bulbs, corms, and tubers will be referred to collectively, using the technical term **geophyte**. This is one of those not-very-user-friendly words (try rephrasing Wordsworth's famous poem about daffodils, "When all at once I saw a crowd, / A host of golden geophytes"), but it makes sense, a good way of describing all those plants which can be

conveniently sold dried in packets at the garden centre. Similar, but not so easily packaged up, are **spring ephemerals**—perennials that have a short spring growth period but die back in summer. Geophytes and spring ephemerals maximise the benefits of a limited season of light, moisture, and nutrients before the competition of surrounding plants and the constraints of the environment reduce these key inputs to plant survival.

Most garden plants are terrestrial: they grow on the ground. In some warmer, humid climates or growing conditions, however, **epiphytes** (plants which grow attached to tree branches or other above-ground structures) may be made use of in the garden; their roots extract enough water from rainfall and atmospheric moisture, and enough nutrients from gathered leaf litter and other debris to survive. **Lithophytes** are similar but are found naturally on rock surfaces.

Most of the structures we call "flowers" in the garden contain both male and female organs, or they may be separate on the plant. A plant where the sexes are on separate plants is referred to as **dioecious**—willows are a good example. In some cases—members of the daisy family (Asteraceae) are the classic example—what we call a flower is in fact a compound structure, containing many very small flowers; these latter are termed **florets**.

Family relationships and genetic origins

I find a very useful key concept in understanding garden plants and their relationships with each other is that of the **gene pool**. Generally when plants are introduced into cultivation it is as the original wild species (or selections of them). Hybridisation often follows, in some cases so quickly that it is difficult to determine which were the original species. The species which can cross with each other form the gene pool. Over time the gene pool tends to expand, as more species are introduced and breeding techniques get more adventurous and sophisticated. In many cases (but of course not always) the genus describes the limit of the gene pool; *Rosa* species all seem to breed with each other, but among *Clematis*, the *C. alpina* types are genetically incompatible with the others. In some families, even intergeneric hybridisation is possible, hence the genetic anarchy of the Orchidaceae.

Where a large genus is both complex and has had a particularly long history in cultivation, divisions have often been carved out: subgenera, sections, subsections, tribes. These are as much to do with our deep human need to make sense of the diversity of life as any "real" divisions. Some of these are botanically based, others are more pragmatic horticultural ones. They are particularly necessary as a kind of housekeeping, an attempt to establish some semblance of order in complex gene pools. These (or the key ones) will be discussed, as they can do much to help make sense of the confusing welter of varieties available to gardeners in some genera.

Occasionally I might refer to some wild plants and cultivars as being **tetraploid**, which means that they have double the normal chromosome count. Tetraploids tend to be larger, and more vigorous. **Polyploidy** is a term which covers doubling and other increases over the normal chromosome count. Not surprisingly, gardeners have always had a bit of a fondness for tetraploids and other polyploids—plants are bigger and often more vigorous, flower colours more intense. Some are like plants on steroids.

Before we leave genetics, a brief word about **mutations**. It is well known that reproduction throws up occasional plants with a random change in their genes which results in a distinctly different plant, those with white flowers being among the most common and widely appreciated. A **somatic mutation**, however, does not involve the genes; an example might be a branch that suddenly starts to develop variegated foliage or flowers in a different colour. Somatic mutations can be maintained only by vegetative propagation (e.g., by division or taking cuttings).

Region of origin and climate

Reference is made to the number of species in the genus and the geographical regions the genus is from. A distinction must be made between **region of origin**, which palae-obotanists are increasingly able to work out, and **centre of diversity**. The latter term refers to areas (which may well not be the same as the region of origin) where a genus has proliferated into the largest number of species; a dramatic example is *Erica*, where a limited number of species are found across Europe and Africa, whereas hundreds occur in the Cape Floristic Region of South Africa.

To a large extent, climate can be inferred from this geographical information, using other reference sources; however, some climatic terms are used. **Continental** climates are those which prevail in the centre of continents or on their eastern edges (at temperate latitudes); hot summers and cold winters, with short transitions, are the norm. **Maritime** climates are those influenced largely by the sea and tend to prevail on west coasts (again, at temperate latitudes); winter and summer temperatures are much more similar—as with the joke that in the south-west of England it can be difficult to tell if it is January or June. **Mediterranean** climates are those with moist, cool winters and hot, dry summers, so plants tend to grow in the winter and be summer dormant. **Montane** climates complicate the picture (although generally, the higher the altitude, the cooler the climate). One side of a mountain chain may receive much more rainfall than the other—such local variations, combined with the geographical barriers that mountains create, mean that some mountain areas have the most incredible botanical diversity.

Temperate climates merge into **subtropical** climates on the eastern coasts of continents where relatively cold winters alternate with very hot and humid summers. Mediterranean climates often merge with **arid** climates, often popularly referred to as deserts, although they may be biologically very rich.

Evolutionary history

Many woody plants have a good presence in the fossil record, and given the knowledge of plate tectonics built up over the last few decades, combined with the evidence of phylogeny, some remarkably accurate evolutionary narratives may be constructed. Herbaceous plants fossilise much less well, however, so for them this story may be absent. The fossil record is now supplemented by DNA analysis, which can extrapolate back and make connections and date origins in a way inconceivable to previous generations. This whole science of evolutionary origins is very new, very complex, very incomplete, and difficult to understand. My references to it will be sparse but tantalising!

What is particularly fascinating here is the way that scientists are reconstructing amazingly detailed narratives of plant evolution—stories played against a backdrop of ever-changing maps. Watching the shape and configuration of the continents develop on one of the many simulations available online is an odd experience: the sheer antiquity of some of the continental landmasses, set against the novelty of some of the movements and conjunctions. As continents temporarily join (perhaps as sea levels fall), plant species can move across and colonise new territory; and as seas divide land masses, plants are isolated and must evolve separately from their relatives.

Fossil evidence of flowering plants dates back to the era of the dinosaurs, the **Jurassic** (201–145 million years ago). (As an aside, it appears that flowers evolved before bees!—so the first flowers were probably pollinated by beetles.) These ancestral flowering plants are thought to have radiated out across the globe during the **Cretaceous** (145–66 million years ago), helped by the fact that the southern supercontinent of Gondwana was breaking up. Basically, if a plant genus had evolved by this time, it would

then be distributed across the globe on the continents. Evolution continued on the now-separated continents through the **Cenozoic** era (formerly the Tertiary). This era is divided into seven epochs.

In the **Paleocene** (66–56 million years ago) the first rainforests, dominated by flowering plants, covered large parts of the earth, and many plant families recognisable today evolved. North America and Asia were intermittently connected by land.

In the **Eocene** (56–34 million years ago) the continents of an extremely warm earth were blanketed with thick forest, with palms almost as far as the Arctic Circle. Grasses evolved during this period but stayed localised. Later the climate cooled and dried out, and deciduous trees began to evolve and spread.

During the **Oligocene** (34–23 million years ago) mountain building began in the west of North America and Europe, but there was a land connection between the two continents, allowing some exchange of flora and fauna. Many plant families (grasses, legumes, sedges) spread widely.

The **Miocene** (23–5.3 million years ago) saw extensive grasslands begin to cover much of the globe. Towards the end of the period, ice ages began to occur. During the mid-Miocene North America and Eurasia were linked via **Beringia**, which is where the Bering Strait is now.

In the **Pliocene** (5.3–2.6 million years ago) South America and North America joined up via the Isthmus of Panama, enabling an exchange of species. Occasional land bridges between Asia and North America allowed further exchanges of species. Grassland and savannah continue to spread, tropical forests to diminish, while deciduous and coniferous forests and tundra covered much of the cooler zones.

Fossils found in northern Europe are often of genera now extinct there but still found in China. There are also plenty of genera with a small number of European/Mediterranean species but with many Chinese (e.g., *Epimedium*). Until the Pliocene, much of Asia and Europe as far as Greenland was covered in evergreen subtropical forest with a temperate mixed or deciduous forest northwards, the fossil record indicating a species mix very similar to that of today's Chinese and Japanese mountains. As colder and drier conditions began to prevail, these belts of vegetation were forced southwards, resulting in regional extinctions, especially in Europe and western Asia. Many plant communities were left stranded. So, for example, the evergreen oak, fir, and cedar forests of North Africa are very similar to those in the Levant or around the Caspian Sea; the species have carried on evolving and are very often different but clearly closely related.

Southern China, with its connection to tropical Southeast Asia, very much acted as a refuge. This region has an immensely long history of continuity, right back into the deep Cenozoic, and many genera may in fact have evolved here. Its conservation is of crucial importance for world biodiversity.

The **Pleistocene** epoch (2.6 million to 11,700 years ago) saw frequent ice ages, with their biggest impact in the northern hemisphere. Plant taxa continually separated and recombined populations as species were driven south or downhill with the cold and then spread back north or uphill again as the climate warmed. Rapid evolution often occurred as a result. It is thought that many herbaceous perennials evolved during this period.

The Pleistocene was followed by the Holocene, although increasingly it is proposed that **Anthropocene** would be a better term, as it was during this period that humanity began to have a major impact on the earth, beginning with the mass extinction of megafauna caused by the hunting activities of our tribal ancestors and continuing with our own present domination of surface geological phenomena, via earth moving, agriculture, and other forms of land management—including, of course, gardening!

Ecology

The growing of garden plants in a more "ecological" fashion became steadily more popular in the latter years of the 20th century, alongside an interest in cultivating species native to a given area. As a consequence gardeners have become much more interested in how plants function in the wild and when "let loose" in garden conditions, as opposed to being kept separated from each other by bare earth, as used to be the case.

Discussion of the ecology of the ancestors of garden plants covers the well-known concept of "habitat" and a related concept, which is perhaps best described as "survival strategy," which I want to focus on here, as I believe it can do a lot to help us understand the longer-term performance of our plants.

Habitat

The words used for habitat are widely understood. **Woodland** is composed, in the main, of **canopy** trees but very often includes **understorey**, which is a usually discontinuous layer of relatively shade-tolerant smaller trees, shrubs, and climbers beneath the canopy. **Woodland edge** habitat is self-explanatory; it encompasses a wide variety of species of both woodland and open country, as well as many shrubs and climbers that tend to occur here to a greater extent than elsewhere. It is a habitat in a perpetual state of transition, generally getting shadier with time and tree growth. Humanity is very good at creating long strips of woodland edge (e.g., along roads, paths, power lines, and fields).

Scrub has much in common with both woodland edge habitat and **chaparral** or **monte,** the dwarf woodland often found in very windy, dry, or exposed habitats. In many regions it is an extremely important habitat, with a mosaic of microhabitats, a bit like a mini-savannah. Unfortunately, it has a rather negative reputation: often impenetrable and too short to provide much good timber, it has little obvious economic value. **Maquis** is similar but generally shorter; this shrub-dominated community replaces forest in much of the deforested Mediterranean region.

Savannah—a complex mix of forest, open grassland, and transitions between the two—is often found in drier temperate regions. Savannah habitats are frequently the result of fire, which sweeps through inconsistently, leaving patches of forest and grassland; hunter-gatherers have traditionally used fire to manage their environment, so extending it well beyond its natural range. There is no doubt we humans find savannah habitats very attractive places; there is an argument that since we evolved in the African savannah our brains are hard-wired to feel at home in savannah, an argument given much force by our strong tendency to recreate it in parkland-type landscapes wherever we go.

Prairie is used specifically for grasslands in the Americas where reduced rainfall (and very often human-generated fire) result in vegetation dominated by grasses and herbaceous plants. **Steppe** is drier still and may be dominated by grasses or subshrubs. Steppe may also refer to the cold winter/dry summer climate of these regions; North Americans will understand this as covering short-grass prairie and sagebrush.

Moorland is used to describe (mostly) upland habitat, without trees, exposed, and dominated by subshrubs and grasses, often with poor acidic soils, sometimes the result of burning. Somewhat annoyingly, there is little agreement among English-speakers on what to call this habitat; for North Americans, pine barrens might be an approximation.

Words for wetlands are numerous: **marsh** is rather general but is specifically dominated by herbaceous plants; fen refers to mineral-rich (and therefore high-nutrient) wet environments; **bog** is acidic and low in nutrients.

Mountains, as alluded to earlier, can be hot spots for evolution. Exploring their flora is

In mountain regions, scree slopes typically become covered in vegetation in time, if moisture levels are high enough. This scree in the Tien Shan range (Kyrgyzstan) is clearly being slowly grown over . . .

always one of the most thrilling of botanical experiences. Truly "alpine" environments are most clearly understood as being above the tree line—however, a great many other plants are grown under this label; indeed, it could be defined as whatever the many societies of dedicated alpine gardeners says it is, including many dry-climate bulbs, slow-growing woodland plants, and, most broadly, anything small and rather choice.

Keeping to our somewhat stricter definition, true **alpines** still include plants from a wide range of habitats, notably mountain grasslands (where perennials thrive but which are too exposed for trees), rocky crevices, and scree slopes. It is these latter that, further

. . . and the new growth will often be tall-herb flora. Here, for example, is the variable purple *Aconitum leucostomum*, one of many perennials that flourish in these very steep locations.

down the mountainside, below the tree line, get particularly interesting, especially for the more general gardener. Scree slopes will eventually get covered in vegetation, which very often takes the form of the so-called **tall-herb flora**, which can also occur in valley bottoms. Poorly understood and often very difficult, or even dangerous, to access, this highly specialised and visually dramatic flora is the home of many of our familiar garden plants—and also, possibly, a seed bed for herbaceous plant evolution. Kept free of trees by avalanches, with a continual flow of highly oxygenated, nutrient-rich underground water, scree slopes offer optimal conditions for maximum plant growth.

The world's **grasslands** have been massively augmented over millennia by human activity, primarily grazing and burning and more recently by a variety of cutting regimes. They are of course dominated by grasses, which have a remarkable ability to survive regular fires and grazing. The richness of their wildflower, i.e., perennial, component will depend partly on the intensity of grazing or the frequency of fires. Humanity has had a particular love of grassland, primarily because of the long history of relying on grazing animals for food and other resources. Consequently many grasslands can be described as anthropogenic (i.e., human-created and -maintained) and without human intervention would turn back into woodland (more on this succession, as it is known, in the next discussion). **Pasture** is grazed, while **meadow** is cut for hay, generally at mid-summer. The latter has evolved, in Europe at least, into a very species-rich habitat, which has probably prolonged the wide distribution of herbaceous species that would have occupied much of the immediate post–ice age world.

Plant survival strategies and succession

Gardeners soon learn that some plants seed themselves around with gay abandon, others spread vigorously, while others grow slowly; lifespans may vary too, with even some trees and shrubs being notably short-lived. Such differences in performance and behaviour are part and parcel of the survival strategies that plants use to maintain the species.

The idea that plant behaviour and performance can be understood and predicted in terms of survival strategies is one which has gained considerably in importance over recent years. Planting styles that aim at creating artificial ecosystems rely very heavily on understanding the concept of survival strategies.

Three strategies are recognised: pioneer, competitive, and stress-tolerating. These are strategies, *not* categories. Most species combine two or three. Nevertheless, some species illustrate a strategy so clearly that they can be spoken of as pioneers or competitors.

Before we go any further, we need to discuss a key concept for plant ecologists, which also helps us make sense of what we do when we garden. But first—a thought experiment. Imagine what happens when a bare patch of soil (such as an abandoned field) gets covered in vegetation. The first year or two is dominated by short-lived weedy species, which produce large quantities of seed—the whole thing a gardener's nightmare. These are the pioneer plants. Later, grasses and longer-lived herbaceous plants take over, forcing the pioneers out. Most of these longer-lived species are competitors, so-called because they endlessly compete for resources.

This process is called **succession** and describes how distinct, and more or less predictable, assemblages of plant species succeed one another over time, before appearing to stabilise. Two further examples of succession: the process whereby solid rock is worn down by erosion and then covered in plants, or the open water of a lake becomes dry land through gradual plant growth combined with sedimentation.

Back to our example. As time goes on, woody plants invade, turning the site first into scrub and then into woodland, initially dominated by short-lived pioneer tree species. Eventually a mature woodland dominated by long-lived trees develops. This appears to be relatively stable, and in the past would have been referred to as a **climax** community, in equilibrium with the climate of the region, and more or less unchanging. This idea of very long-term stability and equilibrium is no longer accepted by ecologists, and the concept of the "balance of nature" is no longer scientifically acceptable. Nevertheless the idea of the climax or **mature** plant community (usually woodland) is a very useful one outside rigorous scientific circles, and I will use it here.

Two species of *Campanula*, drawn by Gertrude Jekyll. The *C. alpestris* (as *C. allionii*) (above) has a dominant root but also new roots growing from a horizontal rhizome, which allows it to spread vegetatively and therefore indicates its clonal character. The *C. thyrsoides* (below) is non-clonal and short-lived; the one point of connection between shoot and root is typical of non-clonal plants. Courtesy of RHS Lindley Library.

Plant longevity

The conventional division of non-woody garden plants into three categories—annual, biennial, and perennial—is grossly simplistic and misleading. An alternative is proposed here, and parallels the divisions drawn with woody plants.

A good place to start is to think about the plants in your garden. Which perennials spread through shoots or runners, or just form slowly spreading clumps? Think now about how you would propagate most of these—in many cases it would be by division, tearing the clumps apart and planting each piece. These **clonal perennials** include the vast majority of garden perennials; they are able to clone themselves, i.e., produce genetically identical copies through various means: spreading rhizomes, roots, or stolons, or simply slowly spreading through short side-shoots which form new roots as they grow outwards. Clonal perennials tend to be competitors.

Annuals and short-lived perennials are overwhelmingly **non-clonal**. They tend to seed rather than spread vegetatively. Needless to say (this being nature, where there are

no hard-and-fast categories), some perennials appear to be non-clonal: they do not form spreading clumps but do appear to live for many years, so heavily do they self-sow. Non-clonal plants are overwhelmingly pioneers, plants that occupy patches of bare ground temporarily in rapidly changing environments. To survive they have to produce seed, which can allow for the next generation to grow and flourish some distance from the parent, in an environment conducive to their growth.

Pioneers

Many popular annuals, biennials, and short-lived perennials are **pioneer** plants, their ability to grow and flower rapidly a great boon to the gardener or parks manager with space to fill, and a need to fill it quickly. They tend to put considerable resources into flowers, seeds, and seedheads, as without doing so, they would risk dying out.

It is also possible to talk about pioneer trees—short-lived taxa, lasting only a few decades (e.g., *Betula* spp.). Typically, they are intolerant of competition, particularly for light, and do not regenerate if cut down.

Pioneer shrubs (e.g., *Lupinus arboreus*, *Cytisus* spp.) are also short-lived. They show no sign of producing new shoots at the base and so have only a limited capacity to regenerate if cut back hard. Some of these may be particularly annoying to the gardener, growing wonderfully for a number of years and then suddenly dying, leaving an empty gap in a border, and (if you are lucky) an evening's worth of firewood. If it is realised that they are short-lived, however, then allowances can be made, and they can be appreciated as the quick space-fillers they are. Subshrubs (e.g., *Lavandula* spp.) tend to have limited lifespans; some may live for more than a decade but can have a long period of senescence, as they age and deteriorate, losing the shape characteristic of the younger plant.

Competitors

For easy gardening, filling space, and reducing maintenance, competitive perennials are good news. Sometimes, however, they can become a little too enthusiastic. **Competitors** seek to maximise available resources and produce as much growth as possible, in order to try to dominate their environment, spread themselves, and smother their neighbours. Consequently they are found in productive environments—those with high levels of light, moisture, and nutrients. Competitive perennials typically aim at cloning themselves, through **rhizomes** (stems which run along or just under the soil surface, rooting as they go), **stolons** (above-ground stems which root at the nodes), or roots which produce shoots when they get near the surface. Each little bit with a root and a shoot is a potentially new plant. Most perennials in cultivation are **clump-forming**, but some send out shoots at random, long distances from the parent—so-called **guerrilla** spread.

Competitors are overwhelmingly long-lived. In theory! In practice, in the garden, a lot depends on their persistence. Perennials vary greatly in the level to which the clonal structure of the plant stays integrated. If older parts of the plant persist over years or are continually replaced in the same space by new growth, we can describe it as **persistent**. If not—if older growth dies back—the clump will break up; in garden conditions, such non-persistent plants have a tendency to disappear, at least from their original location.

Some, essentially non-clonal, non-spreading, perennials may also act as competitors, if they build up a massive underground base and root system. With time, the underground tissue becomes so large with so many buds that it is able to recover from major physical damage, and so is effectively clonal. In some cases, the underground base is very similar to that of the many shrubs, and the plant can almost be thought of as a herbaceous shrub. Some of these are famously slow to establish in the garden—something I will mention where appropriate.

Competitive woody plants dominate mature climax vegetation. A few shrubs run underground to form thickets, very much like perennials; most, however, slowly expand, continually regenerating with new shoots being produced from the base. This woody base can be compared to the clump formed by clonal herbaceous perennials. Long-lived competitive trees have a similar ability to regenerate from the base—cut them down, and they send up masses of new shoots; this is the basis of the traditional practice of coppicing. The underground base of shrubs is often surprisingly large, as anyone who has had the experience of trying to dig one out of a garden will tell you.

Stress-tolerators and stress-avoiders

So far we have discussed plants in terms of their abilities to make the most of the good things in life. But what about when key resources—light, moisture, nutrients—are lacking? This is referred to as stress. Many plants have evolved to maximise the use of limited resources, so that they grow slowly, conserve resources, and protect themselves against predators.

In shade, many species of **woodland perennials** maximise photosynthesis by growing very early in spring and/or by being evergreen. Growth is slow, pollinators can be scarce, and reproduction tends to be by slow vegetative spread.

Subshrubs with grey foliage or leaves reduced to needles or scales minimise desiccation in exposed environments; one of the seeming paradoxes of subshrubs is that they seem to grow in what look to us very different environments: sun-baked hillsides and cold exposed situations. In fact, the stresses of sun and wind amount to pretty much the same thing (desiccation), so their evolutionary survival tricks are likely to be very similar.

Anyone spending time in open, windswept or dry environments will notice that many of the grasses and related plants with linear leaves will form tight tussocks. These have adopted the **cespitose** habit; their new shoots are produced within the sheaths of the previous year's growth, and their root systems recycle the dead leaves that fall around them. These are important for gardeners as, besides being long-lived, they tend to reach a certain size and then stay there.

Stress-tolerant plants face up to stress. There is an alternative, which is perhaps all too familiar to us in our working or even personal lives—**stress-avoidance**! These are plants which are basically competitors while they are growing, but which avoid stress by dying back underground in the stress-inducing season. Many geophytes and some herbaceous perennials do this.

Photosynthesis and plant growth

Most of us are at least vaguely aware of photosynthesis as the means by which green plants achieve the miracle of turning CO_2 and water into sugars using light, so creating the chemical and energy basis for nearly all life on earth. What is less well known is that there are three mechanisms for the chemical pathways involved in the process. The "normal" process, C_3 photosynthesis, uses a three-carbon molecule. The C_4 pathway, which uses a four-carbon molecule, has evolved to make more efficient use of CO_2

(which has tended to decline over geological time) and of water—a great advantage in arid regions. It needs more heat, however, and so is found primarily in tropical genera, as well as in a few genera that have evolved from tropical ancestors and now grow in regions with warm-summer climates (e.g., some North American grasses, *Miscanthus* spp.). Although they make up only 3% of plant species, C_4 plants take up 30% of all the CO_2 fixed by plants.

A third pathway, Crassulacean acid metabolism (CAM), involves keeping the stomata on the leaves shut during daytime and opening them at night to absorb CO_2, storing it as another chemical which can then be converted back into CO_2 in the daytime for normal photosynthesis to work. A very effective mechanism for reducing water loss during the day in arid regions, it is found in many succulents. Both C_4 and CAM have evolved separately, several times over, in flowering plants.

Biodiversity aspects of garden plants

One of the notable aspects of gardening life today is the interest in growing locally native plants. Wild plants form part of a complex web of life, with a series of relationships with other wild plants and animals. When a species is transported to another continent and put into cultivation, it cannot, of course, carry these relationships with it. This is one reason for the widespread interest in growing natives—so that the plants cultivated for ornament may also contribute to local biodiversity. Some relationships with animals may be transported across continents (for example, berries get universally eaten by birds); these are the only ones which will be dealt with, and not in great detail, as there is a considerable literature on the subject.

Toxicity is closely related to the ability of a plant to defend itself against animals set on eating them. There are many Internet sources on this subject (particularly in German, for some reason); in the pages that follow, reference to toxicity will be made only to notable examples. Many plant toxins are **alkaloids**, a class of chemicals which can have therapeutic benefits, if handled carefully.

Plants are occasionally able to poison each other, as part of the competitive struggle for territory. Species that can do this are described as being **allelopathic**.

Among the most adaptable of interspecies relationships is the one with pollinators; it is also one of considerable contemporary concern. Nearly all "natural" (i.e., non-double) flowers can be accessed by bees and other pollinators. Pollinators have preferences, however, and many subtleties are not at all apparent to anyone standing by a flowering plant buzzing with bees (or hummingbirds). No attempt is made to cover these exhaustively, but some indications will be given.

Those who wish to maximise the biodiversity potential of their gardens should research local sources of knowledge, particularly for the often-exclusive relationships between certain plants and the insect larvae which eat them.

Traditions and Uses

Traditional uses of plants are currently enjoying something of a vogue, as part of craft hobbies or the fashionable pursuit of foraging, and new technologies are turning to a wider range of plants for sustainable and environmentally friendly sources of material. Entries will offer an overview of anything significant, giving a flavour of the immensely deep relationships between some plant genera and humanity.

There are a number of sources on plants in folklore and mythology, and a great deal of literature on the herbal uses of plants; but folkloric and mythological discussion of plants is often very vague, and totally different species may occur in different historical, geographical, or ethnic variants of the material in question, so there is little point in going into depth here. The field of ethnobotany and herbal medicine is vast, complex, and of largely specialist interest. Only areas of outstanding interest or contemporary relevance are mentioned. Traditional herbal medicine in particular seems extraordinarily scattergun, as plants seem to be used for radically different treatments from place to place and from tribe to tribe—possibly a reflection of the general futility of prescientific medical practice. Many medicinal herbs are actually toxic at the wrong dose; herbalism has probably killed a good many patients down the ages.

History in Cultivation

There is much detailed material on the early history of many cultivated garden plants (see "Further Reading and Sources") and a great deal on the so-called plant hunters, but very little about what happened to plants when they were established in cultivation—Who raised hybrids? Where? etc. Each entry in this book, whether for genus or plant group, aims to go some way towards filling that gap; at least I hope this will encourage others to undertake more detailed studies, based on primary sources.

A number of historical figures will be referred to by surname only, and a number of historical eras through their names. The following potted history of ornamental plants in cultivation fills in the details. But before we start: the invention of glasshouses and the means of heating them in the late 18th century marked a very distinct breakthrough in gardening history. It makes sense to consider "BG" and "AG"!

Before glass

Ornamental gardening has evolved remarkably few times in human history, arguably only twice (in the Old World at any rate): in ancient Persia and in China. Besides strongly framing the Islamic tradition, the Persian tradition influenced the Roman, which not surprisingly relied heavily on Mediterranean plants, with a major role for aromatic herbs. This body of knowledge was to some extent recovered during the Medieval and Renaissance periods and provided the basis for European garden making.

The first civilisation to really show an interest in plant selection (i.e., picking out good varieties from the wild) appears to have been China's **Tang** dynasty (618–907 AD). During the **Song** dynasty (960–1279), gardening reached a high point in the culture of the scholar-bureaucrats who governed the empire; a wide variety of ornamental species were grown and written about, both as subjects of poetry and in books specifically about plants.

Japan was massively influenced by Chinese culture, starting with the **Nara** period (710–794) when Buddhism became established, the Chinese monks who introduced the new philosophy bringing with them many ideas about gardens, landscape appreciation, and new plants. During the **Heian** period (794–1185), Japan largely turned inward and developed much of its characteristic culture; garden plant selection, independent of Chinese influence, began during this time.

Meanwhile in Europe, after the chaotic period that followed the collapse of Roman civilisation, the **Medieval** period (approximately the 5th to 15th centuries) saw a slow rise in cultural and technological capital. Monasteries were not only focal points of learning but the only locations of gardens, which were very much dominated by medicinal herbs. The writings of German abbess **Hildegard von Bingen** (1098–1179) are a good source for what was available during this period.

Modern historians like to talk about the **early modern period** (c. 1500–1800), which is a way of describing the great breakthrough in history, when Europe—armed with new ideas, knowledge, and (above all) weaponry and sturdy new ship designs—broke out and began to transform the world. There started a steady flow of new plant material into Europe, particularly to the more rapidly developing regions: England, the Low Countries (today's Belgium and the Netherlands), the German-speaking lands, northern France, and northern Italy. Wealthier citizens, either the landed aristocracy or the rising middle classes of the towns, began to become interested in ornamental gardening. Even some artisans started to be able to grow ornamental plants, as we see with the beginning of the **florist** movement.

"Florist" today means someone who works with cut flowers. Until the 19th century, however, it referred to connoisseur growers of flowers. It was not always the wealthy and well connected who brought flowering plants into their first phase of glory, but men (and a few women) of the shopkeeper and skilled artisan class, particularly weavers in Flanders and what is now the Netherlands, during the 16th century. After a phase of religious persecution (many were Protestant), some came to England, where they settled and started off a movement which was to last a good two centuries.

The florists grew flowers in pots and exhibited them, largely for the benefit of other growers. In most cases, shows were competitive and were organised by the societies into which the florists grouped themselves. Until the 19th century, eight categories were recognised in British florist shows: tulip, carnation, pink, auricula, hyacinth, polyanthus, anemone, ranunculus. Most were vegetatively propagated, derived from occasional mutations or sports. Occasionally though someone would sow seed, and a whole new level of variation would be opened up. Prizes were given for the best plants, usually defined as how closely they approached a particular ideal shape. Florists' societies generally flourished in industrial towns—in Paisley in Scotland, it was weavers growing pinks; in Sheffield, it was metal-workers growing auriculas. During the late 19th century the florists' societies generally disappeared, to be replaced by more general gardening associations, and their shows gave way to the bigger and more wide-ranging horticultural shows we are familiar with today. Some florist societies are still going, however, the two oldest being the Ancient Society of York Florists (est. 1768) and the Paisley Florist Society (est. 1782).

A particular feature of the early modern period is the contact with the **Ottoman Empire**. Ruled by Sunni Muslim Turkish-speakers, the Ottoman lands developed a rich garden culture, based in part on the introduction of species from all over the Middle East and central Asia. Sadly, little research has yet been done. Tulips, hyacinths, and irises were the main, but by no means only, imports into Europe.

The early modern period saw much change in nearly all spheres of life, at least in Europe and North America. Towards the end of this period, the rate of change—including the rate of plant introduction—became almost exponential. Ornamental gardening, as opposed to growing plants simply to eat or to treat ailments, took off. Several sources give us a very good picture of the range of species grown and the steady increase in plants of American origin. In England, the **Tudor** period (1485–1603) saw a great increase in prosperity, culture (this was the time of Shakespeare), and gardening. John **Gerard** (c. 1545–1612) wrote *The Herball, or Generall Historie of Plantes* (1597), the definitive book on medicinal herbs, just as herbalism was about to be displaced by aesthetics as the main driver in gardening. John **Parkinson** (1567–1650) wrote *Paradisi in Sole Paradisus Terrestris* (1629), which could be described as the first modern garden book.

The Flemish Carolus **Clusius** (1526–1609) played a crucial role, as early botanist and what we would now call a networker; he played an instrumental role in establishing a botanic garden at Leiden in Holland and at facilitating the distribution of new plants from the Ottoman Empire.

A key text of this time is *Hortus Eystettensis*, the first illustrated book to celebrate garden plants as objects of beauty. Published in 1614 at Eichstatt (in modern-day Bavaria), it is an excellent source for telling us what was then available and regarded as gardenworthy. John **Evelyn** (1620–1706), an English diarist and gardener, is a useful source on the current state of gardening; and Philip **Miller** (1691–1771), chief gardener at London's Chelsea Physic Garden, left us the *Gardeners Dictionary* (1731), which provides an excellent snapshot of what was available to gardeners in Britain. Joseph Pitton de **Tournefort** (1656–1708) often crops up in discussion of this period. He was a French botanist who clarified the definition of the genus, our key concept in making sense of the plant world. He travelled widely, especially in what is now Turkey and Armenia, making many introductions. It is also worth noting that the first cautious experiments in deliberate hybridisation were happening at this time—first in Britain, later in Germany and France.

OPPOSITE Women attending to chrysanthemums, a print produced c. 1890 by Ogata Gekkô (1859–1920), one of the first Japanese artists to win an international audience. Evidence from illustrations suggest that tending to "cult plants" was very much a female activity. Courtesy of the Rijksmuseum, Amsterdam.

European gardeners at this time were frantically trying as many American plants as they could, while the European flora was being exported to the American colonies. In many cases, European imports dominated colonial-era gardens, although Thomas Jefferson and George Washington were two of the first gardeners to actively promote the growing of American native plants. Key figures for introducing Europeans to the North American flora were the English John **Tradescant** the Younger (1608–1662) and later the American John **Bartram** (1699–1777), who exported considerable quantities of seed to British gardeners.

However, while European civilisation was aggressively exploring the world, the Japanese had pulled down the bamboo blinds. During the **Edo** (or Tokugawa) period (1603–1868), the country was effectively closed against the outside world. Under the iron rule of the shoguns, the samurai (who had previously been involved in almost continuous warfare) were controlled for the first time; many of them diverted their energies into cultural pursuits, including gardening. This period saw an explosion of plant selection and hybridisation—even the discovery of the principles of genetics (uncovered later by Mendel), applied at least to *Ipomoea* species. The ornamental plants grown and worked on included both native Japanese and imported Chinese species. The improvement of transportation, which was very marked in the period, and the practice (after 1635) of the alternate-year attendance of the aristocratic landlord class at court in the capital Edo were very effective in the distribution of plants throughout the country. The Edo period could be described as one of the most productive periods yet in ornamental horticulture, anywhere in the world. A particularly vibrant gardening community was based in southern Kyushu, around the **Higo** clan; flower growing here became part of a particular local variant of what we now call *bushido*, the ethical and spiritual discipline of the samurai.

Among the few Europeans who did manage to work in Japan during this period (although only in very restricted circumstances) were the Swede Carl Peter **Thunberg** (1743–1828), who was also crucial in the exploration of South Africa's rich flora, and the Germans Engelbert **Kaempfer** (1751–1816) and Philipp **von Siebold** (1796–1866). They did much to familiarise Western botanists and gardeners with the potential of the Japanese flora and garden culture—and whet their appetite for more.

European powers also faced problems in trying to trade with China. The ruling Ching dynasty, however, was determined to keep foreign traders restricted to small zones in particular ports, so very few were able to travel, which was an enormous frustration to those Europeans who knew that botanical and horticultural riches lay behind the walls

of their compounds. Robert **Fortune** (1812–1880) was one of the few who managed to travel beyond the ports; he brought back many good plants, including the tea bush (*Camellia sinensis*). Others had to beg, bribe, charm, and steal plants from their Chinese contacts, sending plants and seed on long homeward-bound voyages across the oceans; many ships sank, and if they made it, much of their green cargo was dead for lack of care.

A key development at this time was the formation in Britain of the Royal Horticultural Society, in 1804. It played a very important role in developing horticulture and disseminating new plants, growing methods, and information.

After glass

The invention of the glasshouse revolutionised horticulture. It was a gradual invention, however, and had to go hand in hand with the development of heating systems. The 18th century saw the development of the orangery, which enabled the wealthy of northern Europe to overwinter choice plants like citrus fruit and myrtles (*Myrtus communis*). By the end of the century the first true glasshouses had been constructed, along with a crude flue-based heating system, just in time to benefit from the arrival of lots of exciting plants from the Cape of southern Africa. These plants did not mind the dry heat of the primitive heating systems and survived the odd failure; there were crazes for growing Cape heaths (*Erica* spp.), scarcely known today, and geraniums (*Pelargonium* spp.)—still with us!

The nursery industry was an early beneficiary of the new technologies. **Loddiges** of London, which ran from c. 1777 to 1854, were among the first to combine glasshouses with new plant introductions, eagerly taking them in and then propagating and selling them. They were succeeded in the British marketplace by the three Messrs. **Veitch** (fl. 1808–1914). John **Salter** (1798–1874), a cheesemonger turned nurseryman of Hammersmith in London, became an important link between Britain and France, as he ran a nursery at Versailles for many years, later returning home. In France the **Vilmorin** seed company (est. 1743) played an important role.

Glasshouses led to a century-long boom in exotica, which had two distinct aspects: one was the provision of warm-climate plants to wealthy clients, the other the growing of warm-climate plants under cover in the spring, to be set out as **bedding plants** for the summer. From roughly the 1830s, these latter were increasingly cheap, available both for the growing middle classes and for use in public spaces. Over the course of the century, as wealth and access to quality public space for the masses grew, the fruits of the glasshouse became steadily democratised. A very large number of new introductions were trialled as bedding plants, some of which went on to become enormously important commercially; others sank with hardly a trace.

Another aspect of glasshouse-based gardening culture in the late 19th century was the practice of **forcing**. Certain hardy shrubs or perennials would be potted up, brought on quickly in heat, and then sold in markets, to provide early colour and scent for the home. Not all species respond well to this treatment, but those that did were grown and sold in vast quantities.

The glasshouse (or greenhouse) era depended on cheap labour and cheap coal. Its end came as the cost of both rose, most dramatically with the First World War. The writing was long on the wall, however, partly from a growing fashion for a more relaxed style of using hardy plants and partly because of the sensational hardy introductions from eastern Asia, which rose to a green crescendo in the latter half of the 19th century.

The heroic age

The 19th and early 20th centuries were the golden age of the plant hunters—romantic figures who explored eastern Asia (mostly), bringing back an extraordinary wealth of botanical treasure. The region where the borders of Tibet, China, and Burma met was particularly rich—it had never been glaciated and had a vegetation cover with a continous history way back into the Cenozoic. Some plant hunters, such as the French Père Jean Marie **Delavay** (1834–1895), went primarily as missionaries. Later ones—such as the English Ernest Henry **Wilson** (1876–1930), who worked as a collector for the Veitches—tended to be plant hunters, pure and simple. Many of the plant hunters were sponsored by wealthy private individuals with large country estates, e.g., George **Forrest** (1873–1932). Botanical institutions also played an important role in funding expeditions; the Arnold Arboretum (just outside Boston), for example, sponsored Joseph **Rock** (1884–1962).

Russian explorers were meanwhile exploring northeastern Asia. Notable among them was the German-Russian Carl **Maximovich** (1827–1891), employed by the St. Petersburg Botanical Garden. The important role this particular botanic garden played as a conduit for a stream of introductions from the Russian east and other areas of Asia, since its foundation as the Imperial Botanical Garden in 1823, cannot be overestimated.

With the opening of Japan to Western trade in the 1850s and the modernising **Meiji** era (1868–1912), that country's botanical and horticultural treasures were also laid open to discovery. Its nursery industry cashed in too, exporting large quantities of plants to the West.

The sheer volume and quality of what was pouring out of Asia during this time should not blind us to the work of plant hunters in the American West, whose very rich flora was being explored during the 19th century. Most notable among them was one of many Scots, David **Douglas** (1799–1834). South America, however, making a difficult transition from Spanish Empire to fractious and unstable independent countries, attracted less interest, with the plant hunters who did go there tending to be focussed largely on orchids.

The latter part of the 19th century saw the growth of hybridising, turning beautiful but sometimes recalcitrant wild plants into more tractable ones for the garden and public park. Veitch's nursery played a major role in pioneering crossing in many genera, as did Louis **van Houtte** (1810–1876) of Belgium, whose business in continental Europe played a similar role to that of Veitch's in Britain. In France, Victor **Lemoine** (1823–1911) was simply the greatest hybridiser of all time, of perennials, annuals, and shrubs; his influence on our contemporary garden flora is still enormous. He was succeeded by family members who kept the business going until the 1950s. In Germany,

another family business, established by Ernst **Benary** (1819–1893), played a similar role to Lemoine et fils. but is still a major business today. The company established by Georg **Arends** (1863–1952) was particularly notable in breeding perennials through much of the 20th century. Another key German name was Karl **Foerster** (1874–1970), whose role as breeder of perennials was supplemented by his writing. In Britain, the breeder of perennials with perhaps the biggest long-term impact was Amos **Perry** (1841–1913).

Horticulture took off in the United States in the late 19th century, although ornamental plants were always very much second fiddle to the agriculturally useful. Luther **Burbank** (1849–1926) became the best known of all plant breeders in history, through sheer force of personality; a man who pioneered breeding on an industrial scale and yet rejected Mendelian genetics, he always had something of the mountebank about him, and much of his work proved to be without lasting value. With the introduction of investment-heavy technologies, corporations arguably began to play a more important role than individuals, an example being **Burpee** Seeds (est. 1876). This was especially true after World War II, when F_1 hybrids (which allowed for much greater precision in the breeding process) began to be turned out on a large scale.

The 20th century—decline and consolidation

While the 19th-century garden was dominated by artifice—the growing of the exotic in cold climates, the use of very geometric and precisely managed bedding schemes—the 20th saw the steady rise of naturalism. It began with the 1870 publication of *The Wild Garden* by William **Robinson** (1838–1935), who argued for a greater use of wild and naturalised plants. Robinson's journalism and numerous books over the next half century did much to promote hardy perennials and shrubs, as did the garden designs and writings of Gertrude **Jekyll** (1843–1932). By the first decade of the 20th century, the herbaceous border, the shrubbery, the rockery, and even the wild garden were beginning to become almost de rigueur in British gardens. In the United States, an interest in native plants was closely tied to a growing sense of nationhood. In German-speaking Europe, the writings (and plants) of Karl Foerster and of a Czech, Count Ernst **Silva-Tarouca** (1860–1936), were likewise promoting hardy plants and a greater naturalism.

The rock garden was a distinctive feature of the early 20th century in Britain and Germany, a leading figure being the English Reginald **Farrer** (1880–1920), famous for the opinionated eloquence of his often purple prose. This labour-intensive garden form has been in decline, arguably since the 1920s, and a very large number of species are now unobtainable. A notable feature of these years was the rise of the "plantsman," sometimes nursery owners, sometimes private individuals, often writers, always fanatical enthusiasts. The English E. A. **Bowles** (1865–1954) was a fine example, an amateur who added considerably to both botanical and horticultural knowledge.

The period after World War I saw the beginning of a steady, and ultimately massive, loss of warm-climate plants from cultivation, a loss which has only begun to be reversed in the last few decades, as nurseries in the Southeast Asian region began to make their own selections. Hardy plants held up, but there is a distinct sense that a golden age had been left behind. What was supremely successful, however, was a garden style that developed out of the Arts and Crafts movement, largely in England, in the creation of which Robinson and Jekyll were both key. This style, which balances formal frameworking and informal, largely herbaceous, planting, has been immensely successful, long-lived, and very influential well beyond the British Isles.

Germany was the leading gardening nation in the interwar years, with an extremely lively and continual production of new perennials and, to some extent, bulbs. However World War II brought devastation to both German gardening and to a large extent British too. Cultivars continued to disappear during the 1950s, and species were lost to cultivation.

The postwar nursery industry focussed increasingly on plants that could be mass-produced, which has of course given us a great many good cultivars but from a relatively limited number of genera. The development of plants for large-scale public landscaping was also an important development.

The 1960s in some ways marked a nadir in the plant range available, but it was during this time that enthusiasts in Britain, the United States, and Germany began to discover new genera, particularly of perennials, a movement which has greatly gathered strength over the years. The English gardener, nursery owner, and writer Beth **Chatto** (b. 1923) played a crucial role in this, particularly for continuing in the vein of Robinson and Foerster in getting gardeners interested in plants which had not been regarded as gardenworthy in the past. She and the flower arranging movement (which started off her love of gardening) have helped shift the aesthetics of the garden consumer, awakening gardeners to the more subtle beauty of many perennials, as well as sedges and grasses. Also key was nurseryman and writer Alan **Bloom** (1906–2005), who tirelessly promoted perennials.

From approximately the 1990s onwards the most dramatic gains have been in the United States, where an interest in plants, natives in particular, has grown enormously. Eastern European nations too, liberated from communism, and newly emerging economies, have continued to drive the growth of both public and private gardening and variety selection.

GARDEN
FLORA

Abutilon Malvaceae

Abutilon is a genus of around 150 species, comprising small trees, shrubs, and perennials. Its centre of diversity is Brazil, but species also occur elsewhere in South and Central America, as well as in Australia and Asia. The name is from the Arabic (via Medieval Latin) for a member of the genus, or something very similar. A small number of species and hybrids are grown for their ornamental flowers. The most frequently grown group (*A. ×hybridum* and ancestral species) basically offers a yellow-to-red spectrum of flower colour and sometimes attractive leaf variegation. The origin of the cultivated gene pool mostly seems to be Central America and the Southern Cone of South America; in some cases taxa possibly come from high altitude, accounting for their hardiness.

These are plants of open woodland, scrub, and roadsides. All are rapidly growing, and nearly all in cultivation are shrubs. Cultivated abutilons vary in lifespan. *Abutilon vitifolium* and hybrids derived from it tend to be short-lived non-clonal shrubs, a very typical pattern in the family Malvaceae. The *A. ×hybridum* group and similar species are long-lived clonal shrubs, continually regenerating from the base; their habit ranges from stiffly self-supporting to so lax they are best treated as scramblers or climbers. Growth is continuous, and for many in this latter group, flowers tend to be produced continually above a threshold temperature.

Some species have had a wide use in herbal medicine; research indicates there is a real benefit in the treatment of liver ailments. *Abutilon theophrasti*, an annual, has been widely grown for its fibres (for rope and string); it has become an invasive weed in the United States; in parts of Asia the leaves are eaten as a vegetable.

Following their introduction to Europe in the mid-19th century, abutilons became popular conservatory plants, their long-flowering and flexible habit making them ideal for training around pillars and into glasshouse roofs where space was limited. During the 20th century it was found that some were hardy in sheltered spots, taking several degrees of frost. The species hybridise readily, and given the ease with which cuttings can be taken, it is not surprising that some Victorian-era cultivars and hybrids are still in cultivation. Recent decades have seen an increasing rate of hybridisation, with varieties being named on the basis of flower colour and plant habit and shape. Seed is also available commercially, enabling mass displays to be planted in suitable climates.

OPPOSITE A group of abutilons, clockwise from upper left: *Abutilon megapotamicum* 'Variegatum', unknown, *A. pictum* 'Thompsonii', unknown red, *A.* 'Boule de Neige'. Illustration from *Vicks Monthly Magazine*, a gardening journal published in the late 1880s by James Vick of Rochester, New York, which city was an important centre for the nursery trade at the time.

Acanthus Acanthaceae

Acanthus includes around 30 species of shrubby and herbaceous plants from the Mediterranean region, down into northern and eastern Africa and across into western Asia. The name is from the Greek for "spiny"; Acantha was a minor figure in Greek mythology, a nymph who got turned into a plant by Apollo after she fought off one of the unwanted sexual advances of which ancient mythology is so full. African and Asian species tend to be shrubby, not hardy, and often very colourful, with considerable ornamental potential. A handful are in cultivation, some as rather grand border plants in cool temperate gardens but often as a more utilitarian ground cover in warmer climates.

All species seem to be stress-tolerant, able to cope with drought and typically found in open or lightly shaded habitats, including dry hillsides, scrub, and woodland edge. Over the long term, however, they might be best described as competitors, as the root systems of the plants are very large and very resilient once plants are established (which may take years), throwing up suckers from broken roots if the plants are dug out. The hardy herbaceous species illustrate well the typical habit of Mediterranean perennials, growing foliage at low temperatures, and then, if they dry out, becoming summer dormant.

Acanthus is most famous as being the model for the ornamental leaves found on the capitals (pillar tops) of the so-called Corinthian order of ancient Greek architecture. Given the popularity of Classical architecture, the acanthus leaf has reappeared ever since as carved or printed ornament, in both buildings and paintings. The plant had minor uses in herbal medical traditions, Gerard and Parkinson both recommending it for burns and bruises. The rather odd common name of bears' breeches is obscure in origin, without any recorded explanation.

The Romans cultivated acanthus, and the plant reappeared in gardens in the early modern period. It was given a boost by Robinson, who promoted them for their foliage, and has been a mainstay of perennial plantings ever since. Of the several species in cultivation, there has undoubtedly been some hybridisation; plants sold as straight species may not be quite that, but then this is nothing unusual!

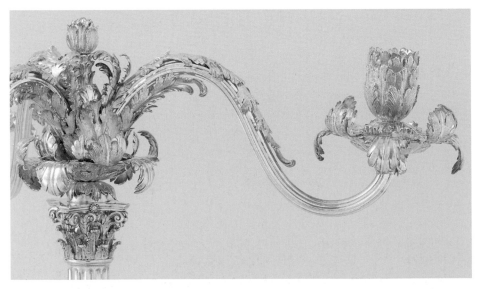

Details of a silver candelabra, made by Johannes Schiotling in Amsterdam in 1772, showing acanthus leaf decoration, a motif which appears time and time again in Western art ever since its use by the ancient Greeks. Courtesy of the Rijksmuseum, Amsterdam.

A plate from "Maples of Japan," a catalogue from the Yokohama Nursery Company, dated 1898. This was one of many nurseries near Tokyo which were producing plants for export during the late 19th century—a time when international plant health controls were minimal. Courtesy of RHS Lindley Library.

Acer Sapindaceae (Aceraceae)

Acer is simply Latin for "maple." The genus contains around 130 species of deciduous trees, found across Eurasia, North Africa, and North America. Southern China is a centre of diversity; it is thought that the genus first evolved in the Yangtze river valley, Jurassic-era fossils being found in central and southern China. The nine North American species evolved at widely different times: eastern American species diverged in the Oligocene and late Miocene, western American species in the late Eocene and mid-Miocene. It is thought that there was early migration to Europe from North America during the Oligocene, before the opening of the Atlantic, but also movement westwards. Many species are grown as landscape trees or for their foliage (which very often colours well in autumn) and their bark. Snakebark species belong to section *Macrantha*; with the exception of the commonly grown *Acer pensylvanicum*, they are east Asian.

> Even the gods in the mountains
> adorn their hair
> with cherry blossom in spring
> and maple leaves in autumn.
>
> (Hitomaro Kakinomoto, 8th-century Japanese poet, quoted in Kashioka and Ogisu 1997)

In nature, maples are found in a wide variety of woodland habitats. A preponderance are relatively small: many species grow in open or montane woodland, or woodland edge habitats. The European *Acer campestre* is a good example; it is never found as a major woodland element but is common in transitional habitats. Many species have a pioneer character and are short-lived, but this is not always so—*A. rubrum* and *A. pseudoplatanus* (sycamore maple) can both seed aggressively and yet be a part of mature woodland. *Acer saccharum* (sugar maple) in eastern North America, is among the few that can dominate climax woodland.

A print by master Japanese printmaker Utagawa Hiroshige (1797–1858) depicting maple foliage as an emblem of autumn. This is one of a series illustrating the seasons. Produced in Tokyo in 1857. Courtesy of the Rijksmuseum, Amsterdam.

Little tradition is attached to maples; the use of the maple leaf as a national symbol in Canada seems to be the most important. Maple syrup can theoretically be produced from all maples in spring but commercially is extracted only from *Acer saccharum*. Originally a Native American practice, it was rapidly adopted by European settlers and is now an important economic activity in the northeastern United States and southeastern Canada. Around 40 litres of sap are tapped from each tree in spring; this is then boiled down to produce about a litre of syrup. The timber is usually hard and good quality; both sycamore and sugar maples are significant timbers. The latter in particular may develop a very decorative patterning, leading to its being popular for furniture and flooring. Finally, as they are among the most colourful of all deciduous trees in autumn, it could be argued that maple leaves have a considerable economic value, in being the basis of a strong seasonal tourism industry in Japan and in the North American northeast, where travelling to admire the leaves is often referred to as "leaf peeping."

North American maples were brought to Europe during the 18th century. Maximovich was responsible for many introductions from the Russian Far East, Japan, and China. Plant hunting in southern and eastern China continued to bring in introductions, until virtually all species had been discovered and introduced by the early 20th century. The larger maples have proved popular as landscape trees, with the numerous smaller species proving to be successful garden plants. The diversity of the Asian species has led to much connoisseur interest in the West.

Maple leaves in *Illustrated Collection of Maples with Classical Poems*, a Japanese book compiled by Ito Thei, dated 1770. Courtesy of Chiba University Library.

East Asian *Acer palmatum* shows particularly high diversity. In Japan the first literary mentions were in the Nara period; it was certainly cultivated in the Heian, when the nobles would hold leaf-hunting competitions in the woods. Over 100 selections were made during the Edo period, with yellow leaves the most highly rated—a Chinese influence, as yellow was seen as the highest-status colour (and traditionally reserved for the emperor); 40 were specifically grown as bonsai. So central are maples to the Japanese autumn aesthetic that the word *momichi*, originally used to describe all autumn colour, came to be a synonym for *kaede*, the original word for maple. Early-18th-century nurseryman Ihee Masatake Itoh played a crucial role in popularising maples, his cultivar names usually based on an expression from a poem he thought appropriate. *Acer palmatum* cultivars began to be exported to the West during the Meiji, with the tree benefiting from the early-20th-century Japanese garden craze.

Achillea Asteraceae

The yarrows are a genus of approximately 85 species of herbaceous perennials. Their feathery leaves are an extremely familiar sight across a wide range of habitats, and their smell is familiar too, herby but bitter and astringent, and reminiscent of *Artemisia* (wormwood), another member of the Anthemideae, one of the tribes into which the vast Asteraceae is divided. Species are found across temperate zone Eurasia, with a few in North America, nearly always in open conditions, with a strong preference for drier habitats; some are a major part of the biomass in central Asian steppe. *Achillea millefolium* in particular has a truly vast distribution; there are at least a dozen different recognised subspecies and local variants of it, which are now particularly confused in North America because of the introduction of plants of European origin.

Yarrows tend to look pretty similar and seem to cross easily, resulting in a growing complex of taxa in cultivation, all grown for their umbel-shaped flower heads, although the silver-grey foliage of some is also valued by gardeners. Plants tend to be long-lived competitors but with short-lived shoots. Consequently they can die out very quickly in less-than-ideal conditions; many garden hybrids have a poor reputation for survival. *Achillea millefolium* is one of the most persistent and has potential as a ground cover plant, even as a lawn substitute.

Achilleas contain thujone, a possibly allelopathic compound. The distinct aroma of all the yarrows indicates an interesting chemistry; they have had very widespread use as a medicinal herb, for a variety of conditions, the treatment of wounds and the reduction of bleeding the most prominent—hence a popular herb with army herbalists across many cultures and the logic behind the genus name: Achilles was a legendary Greek warrior hero. There is evidence that yarrow is effective in cases of heavy menstrual bleeding and against high blood pressure. Yarrow tea is a popular cold cure. Surprisingly little research, however, has been done on its pharmacology.

There is anecdotal evidence that yarrow has mildly hallucinogenic effects, supported by its use for divination; stalks of a yarrow species were originally used by practitioners of *I Ching*, the key Chinese divination text. In parts of Britain and the German-speaking lands, yarrows were used against witchcraft and in the British Isles played a part in a variety of love charms. *Achillea millefolium* was occasionally used as a bitter addition to beer, with suggestions in some cultures that it somehow enhanced intoxication.

Colour forms of *Achillea millefolium* have been picked out by gardeners for centuries. From the 1900s onwards, British growers began to make crosses between this and several other species, resulting in complex hybrids where parentages are often obscure. Wilhelm Kikillus bred the Galaxy Hybrids in Germany in the 1980s. Ernst Pagels also produced seven varieties during this period, based on *A. millefolium* and *A. filipendulina*, a species introduced from the Caucasus in the early 19th century; the best known, 'Walter Funcke', is named for a landscape architect who was persecuted by the Nazi regime and went on to become a prominent landscape designer in communist East Germany. These modern hybrids are popular cut flowers, partly because they fade beautifully as they die.

Aconitum Ranunculaceae

Around 250 species of herbaceous perennials, the monkshoods, or aconites, are thought to derive their genus name from the Greek for "without struggle," referring to the quick death afforded by this plant's notorious toxicity. In origin they are overwhelmingly Eurasian, with a few species in North America, almost entirely west of the Rockies. East Asia would appear to be a centre of diversity. With tall spikes of blue-violet flowers, the "queen of poisons" is a popular herbaceous garden plant for summer colour in the border. For a plant that is so toxic, cases of accidental poisoning are very rare—there is nothing remotely edible-looking about the plant, although poisoning can occur through prolonged handling of roots or foliage. Livestock avoid the plant, much to its advantage in the wild, where it escapes grazing.

Aconitum napellus growing in limestone pavement in the Italian Alps with *Prenanthes purpurea*.

The taxonomy of this complex genus is still much discussed and perhaps will remain forever unresolved. The evolution of monkshoods illustrates what often happens to species that grow in mountainous areas, as climate shifts over millennia alternately push plants into genetic isolation (so enabling new races or even new species to evolve) and then into proximity (enabling hybridisation). Monkshoods are remarkably consistent in overall form, range of colours, and broad ecological preference, with garden varieties being selected on the basis of colour and good performance. One characteristic is particularly interesting: the ability of some east Asian species (e.g., *Aconitum volubile*) to climb; some others grow as "normal" upright perennials, but if they encounter overhead obstruction, will start to bend, enabling them to reach through tree branches—a clear example of evolution in action, for the plant is clearly at the very beginning of becoming a climber.

Monkshoods are found in woodland edges, mountain meadows, and tall-herb flora, the common factor being a cool root run and plentiful moisture and nutrients during the growing season. There is little tolerance of drought, and they readily become summer dormant if dry. Many are found at higher altitudes. They are clonal perennials, clump-forming and spreading only slowly. Some east Asian species have roots that form clusters of annual tuber-like structures rather than an interconnected and persistent root mass. Flowers are helmet-shaped—plants are known as "helmets" (*Eisenhut*) in German. The flower shape has evolved to be accessible only to larger bumblebees; other bees, however, frequently bite their way into the rear of the flower to access nectar.

Wherever monkshoods/aconites are found, the lore concerning them is dominated by their qualities as a poison and only secondly by their medicinal value. Aconite has a long history of use as a poison across its geographical range: for killing vermin (hence wolfsbane, another common name), executing criminals, tipping arrows in warfare, and of course for murder. Death usually occurs from asphyxiation as a result of paralysis of the nervous system, although the cocktail of alkaloids the plant contains also act on the cardiovascular and digestive systems. The plant frequently occurs in mythology and literature as a herbal poison, from the ancient Greeks to Harry Potter.

To be used as a medicinal herb, the plant material needs treatment to reduce the level of toxicity, something that was achieved in the past only by sophisticated medical systems such as the Tibetan, Chinese, and Ayurveda. In central Asia the root tubers have been eaten but only after boiling for over 10 hours to destroy the toxins. Modern medical techniques enable the complex range of chemicals in monkshood to be differentiated and separated enough for the plant to be used medically, even to the extent that wild-collecting is beginning to have an impact on natural populations in China and India, and almost certainly other countries. Modern uses focus on monkshood's antimicrobial properties and ability to inhibit enzyme action.

> Many taxonomical problems arise from the notorious hybridity within the genus and subsequent origin of putative genetic hybrids, which form a morphological continuum between the parental species.
>
> (Novikoff and Mitka 2011)

Although grown in herbalists' gardens, by the late 19th century monkshoods were more likely to be appreciated as ornamental plants. European and some Asian species formed the bulk of the older gene pool. Arends did some breeding, as did some British nurseries during the 20th century, but hybrids have tended to be simple. Recent collecting in China, Japan and Korea has hugely increased the gene pool in cultivation, especially of semi-climbing and later-flowering species.

Actaea <small>Ranunculaceae</small>

The name derives from the Greek/Latin for elder (*Sambucus*), after the similarity of the leaves. Researchers have proposed combining the eight-species-strong *Actaea* with the 15 or so species of the related *Cimicifuga*, a decision based on DNA evidence and a statistical analysis of minute morphological characters (Compton et al. 1998). Interestingly, popular culture recognises the similarity, as cohosh is often used in North America as a common name for both genera. Widely distributed across the northern hemisphere, largely in cooler or montane temperate climates, these two genera of herbaceous perennials share much in common but are superficially different, (old) *Actaea* being shorter and bearing berries, *Cimicifuga*, taller with dry pod-like seedheads.

All species flourish in cool, moist soils in shade or light shade. So long as requirements for moisture and high humus content are met, the range of habitats plants can grow in is wide, and they are not the slow-growing specialists often seen in woodland conditions. They are long-lived, and with their dense root network they compete effectively and can dominate their environment; they tend not to spread, however, and stay as tight clumps. *Actaea* flowers are primarily pollinated by beetles but also visited by flies.

Well known as poisonous, many species have nevertheless been used in herbal medicine by Native Americans and settlers, especially in the Appalachians. The range of conditions treated is wide but seems to focus on women's issues and secondly on nervous and neurological conditions. *Actaea racemosa* (aka black cohosh and black snakeroot, among myriad other names) is still highly rated by herbalists, with preparations for the management of menopausal symptoms extensively sold by the health supplement industry. Research does not support a clear position on its efficacy, and it is suggested that more work needs to be done. *Actaea rubra* is the most important herb in a variety of herbal bundles used by shamans in the North American Cheyenne culture.

Plants have been in cultivation since the 19th century but did not become particularly popular until the end of the 20th century. One factor which has focussed interest on the entire genus has been the occasional appearance of dark forms of the east Asian *Actaea simplex*, now classified as the Atropurpurea Group, the naming of which recognises the genetic diversity of a range of plants often propagated

Actaea racemosa, growing in woodland, North Carolina Appalachians.

by seed. The first appeared at a Surrey (England) nursery in 1932 (and was named after it—'Elstead'), but fame had to wait until the 1990s, when several other similar mutant forms began to appear as the scale of commercial production of the genus increased.

Agapanthus Amaryllidaceae

The name is from the Greek, *agape* ("love") and *anthos* ("flower"). The genus is dogged by nomenclatural confusion: there are six to 10 species, depending on which botanist you believe. All are herbaceous perennials, instantly recognisable for their strap-shaped leaves and round heads of blue flowers. They are native and restricted to southern Africa, found across a range of the region's climate zones, mostly cooler and montane. Parkinson described the type as *Narcissus marinus exoticus*, after its introduction from the South African Cape in 1629. Others described it as a hyacinth. Botanists have put the genus in the Liliaceae, Alliaceae, Themidaceae, its own family Agapanthaceae, and a revamped Amaryllidaceae, now supported by DNA evidence. Horticulturally, two divisions are useful to recognise: deciduous species (from areas with only summer rainfall) and evergreen species (from areas with more evenly distributed rain).

Preferring moister situations in rocky grasslands, agapanthus are competitive clonal perennials forming extensive and very long-lived spreading clumps that can exclude all other vegetation, making them useful as minimal-maintenance ground cover in warm-climate zones but also a dangerously problematic invasive species. Flowers are pollinated by bees and sunbirds, the latter preferring pendant flowers, such as those of *Agapanthus inapertus*.

The leaves, sap, and rhizomes of *Agapanthus praecox* are highly toxic to humans and may cause ulceration and skin rashes, particularly on children. Nevertheless, extracts are used as herbal medicine in several African traditions, particularly in promoting easy childbirth.

Introduced to England in 1629, when Parkinson records it, *Agapanthus* is not heard from again until half a century later, when Dutch occupation of the South African Cape ensured that more plants were sent back to Europe. It was grown in pots and overwintered inside until the early 20th century, when a few more-adventurous gardeners in the west of the British Isles began planting it outside. Widely distributed through the European empires in the 19th and 20th centuries, *A. praecox* has proved extremely successful as a landscape and garden plant across suitable climate zones, primarily Mediterranean and subtropical. Nearly all evergreen hybrids are descended from it. *Agapanthus inapertus*, with narrower, drooping flowers and deep blues, is increasingly important in breeding work, however. Australian breeding is trying to produce cultivars with low seeding potential, which would minimise the danger of escape into the wild. Breeding in Britain from a wide gene pool has now resulted in many hybrids that can be regarded as safely hardy.

"The Blue African Lily (*Agapanthus umbellatus*) [now *A. africanus*] at Eastbury Manor, Surrey" says the caption in a copy of *Gardening Illustrated* from 1887. Grown in tubs so they could be brought indoors over the winter, agapanthus are still widely used in this way, although many are hardier than originally thought.

Alcea Malvaceae

The name is from the Greek for a type of mallow. The hollyhocks, a genus of around 60 species from across Eurasia, are familiar as informal garden plants with a long history in cultivation. All are technically herbaceous perennials but with a tendency to have such tough, fibrous, upright stems that an evolutionary transition to woody shrubs can easily be imagined. Drier or transitional habitats are preferred. These are short-lived non-clonal pioneer plants, as can be appreciated by the alacrity with which the cottage garden hollyhock grows in paving, different colours often replacing each other from year to year. The flowers are popular with bees but also with hummingbirds in the Americas.

At the Aoi Matsuri, a major festival held in Kyoto every May, hollyhock leaves are used as a decorative and symbolic element (the plant was held to protect against storms); the celebration has been held annually since

OPPOSITE Double hollyhocks from *Hortus Eystettensis* (1614), along with some single-flowered varieties.

I

Malva hortensis flore simplici albo.

II

Malva hortensis flxe simplici incarnato.

III.

Malva hortensis flxe simplici rubro.

LIII.

Malva hortensis flxe plens atrorubente.

V.

Malva hortensis flore pleno rubro.

Georg Mack.

the 6th century, one of the longest continual runs in the world. The leaves (not the flowers) were used as the personal badge of the Tokugawa family, who established the Edo period, a crucial era in Japanese history. Founder Tokugawa Ieyasu insisted on using his hollyhock emblem and not accepting the imperial chrysanthemum as a way of showing his support for the shogunate as opposed to giving political power to the emperor.

Like most mallows, hollyhock's mucilaginous qualities make it soothing for herb tea preparations and poultices; its traditional herbal uses have been extensive. Both leaves and flowers are edible but tasteless. The stems have been used as a source of fibre by country people, and in the early 19th century, there was an attempt to develop it as a commercial crop in Wales. Several hundred acres were planted up—unsuccessfully but no doubt beautifully!

Alcea rosea seems to have become a domestic plant during the Medieval period, as various colour forms can be seen in 15th-century German paintings; they were possibly introduced from cultivation in the Islamic world. The early 17th century saw the introduction of *A. ficifolia* from Russia, bringing in genes for yellow flowers and more vigorous plants. By the 18th century the hollyhock had become a cottage garden plant across Europe. Goethe grew an allée of them and even had tea parties in their honour.

Doubles were known centuries ago, with *Hortus Eystettensis* (1614) displaying some particularly voluptuous examples. With such a wide range of colour and petal arrangements, the flower became something of a staple in mid-19th-century flower shows. William Chater, a commercial grower in Essex, perfected a relatively stable double strain in the 1880s. Scores of cultivars were developed, propagated by cuttings; but rust struck at the end of the century and wiped them all out, forcing gardeners to treat hollyhocks strictly as biennials, always grown from seed.

Alchemilla Rosaceae

Alchemilla is a genus of around 300 species from the cool temperate to subarctic zones of Eurasia and montane regions in the Americas and Africa. The flowers are often self-fertilising, resulting in a large number of micro-species with minimal differences between them. The name comes from the Arabic for "alchemy," as the way that water droplets sit on the leaves, mercury-like, was seen as almost magical. It was believed in Medieval Europe that this water was the purest form there was and so ideal for alchemical preparations. The phenomenon is actually the result of microscopic hairs on the leaves that trap air, creating a hydrophobic surface.

Alchemilla species are found in habitats ranging from woodland edge to open but nearly always with relatively high levels of moisture. Cold conditions and short growing seasons appear to be tolerated.

Alchemilla mollis in the wild (old pasture in Powys, Wales), illustrating perhaps why it took so long for the plant to be considered an ornamental. In garden conditions it becomes far more conspicuous.

Plants are clonal clump-formers, with considerable capacity to seed, showing a competitive-pioneer character. The tight clumping habit of the leaves seen in cultivation is not typical—when the plants are growing in meadow and other dense habitats, their leaves and stems are usually intimately mingled with surrounding vegetation.

The shape of the leaves of many common European species brings to mind a spread cloak—hence the old Catholic name, Our Lady's mantle. Protestantism scrapped the "Our," and in any case the Marian connection was a Christianising of an older Germanic pagan link with the goddess Frigg. As a herb, lady's mantle's astringent qualities have been used to reduce bleeding wounds and excessive menstruation.

With their frothy, pale yellow-green flowers, *Alchemilla* species have largely been disregarded: Robinson does not mention them, and Silva-Tarouca gives them only a minor entry. Only in the later 20th century were their subtleties appreciated. *Alchemilla mollis* is now one of the most common British garden perennials. A small number of other low-growing species are also grown as rockery or edging plants.

Allium Amaryllidaceae

The number of *Allium* species (the name means "garlic" in Latin) recognised by botanists varies greatly, with 750 something of an average figure. All are geophytes. A key characteristic is the distinctive chemistry which gives "garlics" their onion smell: it's all down to sulphur derivatives of the amino acid cysteine. The genus has a centre of diversity in central Asia, but it is widespread across the northern hemisphere, with a few in Africa and South America, in Mediterranean, temperate, and montane climate zones. Plants are found in a wide variety of habitats, with a tendency towards the dry. Many are bulbous and able to cope with severe seasonal drought. A few are woodland or even wetland species. Those with bulbs include many spring ephemerals from seasonally dry habitats. Those from moister habitats with a longer growing season are more likely to be rhizomatous and in some cases flower later in the season. Many species seed and spread quickly, helped by the fact that the period between germination and flowering is much shorter than with most bulbs, sometimes only a year; *A. triquetrum* is one which has become particularly problematic as an invasive alien. All are banned from New Zealand.

> Across the northern hemisphere, probably since the first hot meal was prepared, alliums have provided the necessary flavour that turns plain fare into gourmet living.
>
> (Davies 1992)

Garlic has been widely used in a range of herbal treatments. Evidence that it can be efficacious is relatively good, for cold prevention, repelling biting insects, reducing cholesterol, and helping prevent heart attacks. Alliums were widely used in charms (e.g., the well-known aversion that vampires have to garlic), but their culinary uses were strongly cultural, the Elizabethan English, for example, not using them either in food or in herbalism. Gerard was particularly scathing in his *Herball* (1597): "As for repeating of foolish and vaine figments, the coniuring of Witches, and magicians' enchantments, which have been attributed unto those herbes, I leave them to such as had rather plaie with shadowes, than bestow their wits about profitable and serious matters." Nevertheless, many species have been eaten on a regular basis—indeed, probably every allium is edible. Of those in cultivation as ornamentals, however, only *Allium cernuum* was a major food item. It was once believed the name "Chicago" derived from the name for the plant in a local language; it is now believed to be derived from *A. tricoccum*. The honey was also apparently very good.

Around 100 of the species are in cultivation, most of them only among collectors. Keeping track of which species was cultivated where and when is difficult. "Moly," for example, meant any *Allium* species to the Elizabethans. Parkinson said he grew 14 species, but there was little interest beyond the culinary and herbal for some time. Most alliums in cultivation were introduced relatively late, an example being *A. cristophii*, introduced by the Dutch bulb company Van Tubergen in 1930. This and the other "drumstick" species of central Asia took a surprisingly long time to make an impact, but by the 1990s they began to take off, and nowadays they have become almost a cliché at the Chelsea Flower Show.

Smaller alliums are primarily grown as rock garden plants, but their seeding tends to limit interest. This aspect has however proven useful in the new technology of green roofs, as many of them flourish and spread in shallow substrates.

Amaranthus Amaranthaceae

Amaranthus is a genus of around 60 herbaceous plants; the name is from the Greek for "unfading," a reference to the similar colour and appearance of the tiny flowers and the seedheads, both of which are packed into a distinctive head. They are found throughout the world's frost-free climates. These are annuals or at the most short-lived perennials, with a clear pioneer strategy and therefore usually encountered as arable weeds or on waste ground; they are huge beneficiaries of the human impact on warm-climate landscapes.

Amaranth species were some of the most widely grown crops in Pre-Columbian Central America, the seed being both eaten as a staple and as an ingredient in a range of drinks and preparations, often linked to festivals. Many of these uses survive in contemporary Mexican cuisine, although links to the indigenous religious practices of the region made the plant a target for Spanish Catholic intolerance, and its use declined considerably after the conquest. One particular reason for the ire of the Christians was that images of amaranth and honey were commonly sculpted in festivals and used as a focus of prayer.

Amaranth seed is extremely nutritious, with around 30% more protein than true grains like wheat or rice. This, and the fact that it is gluten-free and very easy to grow, has resulted in a great recent increase in interest in the crop, with old varieties being rescued and used in breeding programmes. Some strains contain saponins and other chemicals that reduce digestibility, but these can be dealt with in processing. The leaves too are nutritious, easy to cook, and have good keeping qualities; their use as a vegetable (e.g., the callaloo of Caribbean cuisine) is almost universal wherever the plants grow naturally. The roots are also edible. A dye may be extracted from the flowers, and the plant has given its name to an artificial dyestuff.

Although known to the ancient Greeks (the plant was sacred to the goddess Artemis), amaranths were not cultivated until the 16th century, when *Amaranthus caudatus* (in French, the "nun's scourge") began to be used as an ornamental, especially forms with purple leaves and long drooping flower clusters. During the 18th century it, and possibly others, became relatively popular across Europe. During the 19th century the exotic shapes and colours of South American species such as *A. hypochondriacus*, with brightly coloured branching heads, were seen as perfect additions to the tropically exuberant planting styles of the era. The 20th century saw a decline, but there has been something of a revival recently, with much new selection for leaf and flower/seedhead colour.

Anemone Ranunculaceae

Meaning "daughter of the wind" in Greek (although there are other theories about the origin of the name, including one involving the Sumerian god of vegetation, whose name in Phoenician is very similar), *Anemone* comprises some 120 species, all herbaceous perennials. It formerly included *Hepatica* and *Pulsatilla*, which were hived off separately in the latter part of the 20th century. There is a wide range of form, including many geophytes with tubers and rhizomatous species. *Anemone* is overwhelmingly temperate northern hemisphere, but with some species to be found in cooler and montane areas in Africa, Australasia, and South America. There is a clear split in climate preference between cooler temperate species and those of Mediterranean to semi-arid climates. Molecular analysis reveals that there are two clear groupings: one is the Asian woodland group (*A. hupehensis*, *A. vitifolia*, etc.) and the other includes all the tuberous species (*A. coronaria*, *A. blanda*, etc.) and rhizomatous woodlanders like *A. nemorosa*. The latter group, however, have a very disjunct distribution. The same study (Hoot et al. 1994) also suggests that *Hepatica* and *Pulsatilla* rejoin *Anemone*. The yellow-flowered

Anemone pavonina on Poros Island, Greece.

"Five Studies of Anemones" by an anonymous draughtsman, c. 1760–70. Two of the flowers exhibit a much more prominent central boss than is ever seen in modern varieties. Oil on paper, believed to be Dutch. Courtesy of the Rijksmuseum, Amsterdam.

*flore pleno.

Described as "Single and Double French Anemones" in the caption, this engraving illustrates the artistic licence used since time immemorial to enhance the size and fulness of flowers. From *Gardening Illustrated*, 1881.

A. multifida has a particularly disjunct distribution, occurring in locations from the Canadian Yukon to Argentina; it appears to have spread along mountain ranges during the Pleistocene.

Anemone species inhabit a varied range of habitats: tundra, grassland, woodland, semi-desert. All are potentially long-lived, although the tuberous species from seasonally dry habitats, such as the familiar *A. coronaria* types, will not persist in moist climates; in the wild these are spring ephemerals, frequent as weeds of cultivated areas or in deforested habitats. The woodland species can be slow to establish but are very persistent later, with a long-term competitive dominance of the forest floor—the well-known Japanese anemones particularly so. Woodland species vary in their performance: the smaller American and European ones such as *A. nemorosa* tend to have a short spring growing season, while the Asian ones tend to exploit the potential of the entire growing season, and only flower late.

The flower gets mentioned in a few Classical texts, e.g., as the nymph Anemone, or the flowers growing from the spilt blood of Adonis; these of course refer to the blood red species of the eastern Mediterranean. Little other lore is recorded.

Anemone coronaria came to late Medieval Italy in soil imported as ship's ballast from the Holy Land. Spread across the Campo Santo in Pisa, it grew flowers, blood red in colour, which were regarded as miraculous; seed was then distributed all over Europe by pilgrims. Crossed with *A. pavonina* to produce *A. ×fulgens*, a rich gene pool resulted: over 50 varieties were bred in France and Italy and were available in northern Europe during the 17th century. Further contributions to the gene pool came from the Ottoman Empire and possibly elsewhere around the Mediterranean. The Dutch soon took up commercial cultivation, and the flower became popular with florists. The widely cultivated de Caen and St. Brigid strains were developed by growers in Normandy in the 18th century; they remain popular in the cut flower trade.

The first so-called Japanese anemones were introduced by Fortune in 1844; they were actually of Chinese origin but had been grown in Japan for centuries. 'Honorine Jobert' was a remarkably early (1851) somatic mutation of *Anemone ×hybrida*; named for the daughter of a nurseryman in Verdun-sur-Meuse, France, where it first occurred, it is still the finest and most robust variety. Another contribution to the gene pool, *A. hupehensis*, was introduced in 1908, from China. Further selection has proceeded at a very modest pace ever since, with the first dwarf cultivars, raised in the United States in the early 2000s, the most dramatic recent development.

Antirrhinum Plantaginaceae (Scrophulariaceae)

The name is from the Greek for "like a nose," a reference to the flower shape. The 23 species occur in central Europe southwards, North Africa, and the United States, although there is much dispute about whether the American ones should be included in the genus. Snapdragons, as they are often known, are generally short-lived pioneer perennials and subshrubs of rocky places. Most, including the common *Antirrhinum majus*, produce very weakly woody overwintering growth. In the more arid Mediterranean climate zones, however, they may function as winter annuals. The characteristic shape of the flower has evolved to favour pollination by bumblebees, which have the necessary weight to pull down the lip of the flower and gain access.

Plants have been used medicinally in folk herbal traditions, and indeed, recent research indicates potentially useful antimicrobial action. The plant has a long history of use in genetic research as a model organism.

Well known since the Medieval period, *Antirrhinum majus* has often naturalised on old walls throughout Europe. It was certainly in cultivation by the 16th century, and five colours are represented in *Hortus Eystettensis* (1614). By the beginning of the 19th century, many more colours and doubles had appeared. During this century, intensive breeding resulted in yet more colours, bi-coloured and striped varieties, variously shaped flowers, and perhaps most importantly, plants with different heights. Some modern seed strains will produce plants that grow to 90cm, others to only 20cm, enabling them to be used for very different kinds of display. In Victorian times, some varieties were reputed to grow up to 2m, to form great shrub-like mounds covered in flower. The plants are popular and have become naturalised in India, where they are known as dog flowers, after the flower shape.

"Haines' New Rust Proof Snapdragons" announces the cover of the Charlotte M. Haines ("Seeds Grown by a Woman") catalogue of 1936, from left to right: 'Indian Girl', 'Scarlet Defiance', 'Buttercup', 'Daintiness'. The company was based in Rockford, Illinois.

Aquilegia Ranunculaceae

The name is probably derived from the Latin for "water carrier," referring to the nectar in the base of the spur, although an association with the Latin for "eagle" (*aquila*) is also a popular suggestion. There are around 60 species, with centres of diversity in the Balkans and in high-altitude areas of the southwestern United States and neighbouring Mexico. Plants usually act like non-clonal short-lived perennials, with a definite pioneer character, and are freely seeding. Favoured habitats are very much cooler climatic regions, with some alpine species, but most are found in open mountain woodland and transitional habitats. Ancestral species are thought to have originated in eastern Asia and then spread into North America via the Bering Land Bridge; there, they diversified rapidly, in as little as 40,000 years. The Eurasian species are generally blue/purple and pollinated by flies and bees; in North America hawk moths pollinate species with the paler flowers, bees the blue/purple ones, and hummingbirds the red ones. It is thought that colour and spur length evolved relatively recently to attract particular pollinators, in a manner similar to the link between insects and flowers to be seen in orchids.

Aquilegias have been used in herbalism, both in Europe and by Native Americans, particularly against jaundice, but Linnaeus mentions deaths caused by overdosing—they contain toxins similar to those present in the related *Aconitum* and *Actaea*.

Doubles of *Aquilegia vulgaris* were known by Gerard, while Parkinson noted the tendency of the species and selections to constantly cross-pollinate. The late 19th century saw crossing with the many newly introduced species from North America, such as *A. longissima*, *A. coerulea*, *A. canadensis*, *A. formosa*, and *A. chrysantha*. Breeding seed strains has continued since, with a focus on long-spurred flowers with enough lasting power to be good cut flowers. Smaller (often higher-altitude Japanese) species have also been selected from, for rock garden use.

Aquilegia vulgaris in a bottle with peonies, irises, and *Convolvulus tricolor*. Oil on paper by Dirck de Bray, dated 2 June 1674. Dutch. Courtesy of the Rijksmuseum, Amsterdam.

Asters Asteraceae

Asa Gray might well have been confused (see sidebar), as the genus *Aster* (whose name derives from the Greek for "star") contained around 600 species when it was defined morphologically. But modern genetic analysis has revealed the real relationships underneath an umbrella of basic shared floral architecture. *Aster* is now correctly restricted to the Old World species, found across temperate Eurasia, with New World species reclassified into no less than 10 new genera, though all are treated within the tribe Astereae; of these, the following are the most horticulturally relevant:

- *Doellingeria*. Includes *D. umbellata* (the former *Aster umbellatus*).
- *Eurybia*. Includes the now widely available *E. divaricata* (formerly *Aster divaricatus*).
- *Symphyotrichum*. A genus of 90 species, including the vast majority of the American asters grown in gardens.

> Never was there so rascally a genus ... My wife now excuses me to her friends for outbreaks of ill-humor, the excuse being that I am at present in the valley of the shadow of the Asters.
>
> (Gray 1894)

The species whose descendants populate our gardens are all clonal perennials, although the level of clonality varies considerably, from the very dense tufts of *Symphyotrichum puniceum* to the running roots of *S. novi-belgii*. Typically they are competitive-pioneers of productive and often transitional environments, often inhabiting woodland edge or wetland habitats or the earlier successional stages of prairie

Symphyotrichum novae-angliae with *S. cordifolium* growing alongside a *Solidago* species in woodland edge habitat, Ragstone Prairie, Illinois.

PRINCESS MARIE LOUISE

MISS MUFFETT

HILDA BALLARD

FLOWERS NATURAL SIZE AND COLOUR

Three novi-belgii varieties from the 1938–39 Old Court Nursery catalogue. Picture by Ernest Ballard, from left to right: 'Princess Marie Louise', 'Miss Muffett', and 'Hilda Ballard'. Courtesy of Paul Picton.

grassland. They tend to flower late, with rapidly maturing seed, and can quickly spread. *Symphyotrichum novi-belgii* has naturalised in Europe but rarely problematically. All asters are important for late-season pollinators, especially butterflies and bees, although research does suggest that the nectar offers quantity rather than nutritional quality.

Asters have had very little use or significance in traditional cultures, the appreciation of their flowers being a thoroughly modern interest. The Hungarian revolution of 31 October 1918 became known as the Aster Revolution, as protesters wore the flower.

Of the Eurasian, "true" asters, *Aster amellus* has been in cultivation from the end of the Medieval period onwards; it participated in the early-20th-century "aster boom," with many selections made and a cross with the Himalayan *A. thomsonii*, to produce an excellent hybrid, *A. ×frikartii*.

Symphyotrichum novi-belgii was found in what is today New York State in 1687. This and many other species were introduced in the early 18th century, gradually spreading as garden plants across Europe. It was a common garden plant in Germany by the mid-19th century and already naturalising. The genus as a whole attracted a lot of interest in the country, with the Berlin Botanic Garden having over 50 American species in 1812; of these, many were soon made commercially available. Some private collections in Britain were very extensive too, although interest in them was not widespread until they were promoted by Robinson in the latter part of the century.

A friend of Robinson's, Ernest Ballard of Herefordshire, was the first British nurseryman to specialise in asters; he raised a large number of varieties, mainly of *Symphyotrichum novi-belgii*, many being contributed (as *Aster novi-belgii*) to an RHS trial held in 1907 that included around 300 cultivars. Many more were bred by Amos Perry, who also worked with *S. novae-angliae*. German breeders too were very productive from the early years of the century until World War II.

Two pictures from the 1950s taken at the nursery of the leading British breeder of asters, Percy Picton, of Old Court Nurseries, in Colwall, Herefordshire. The business is now run by his son Paul and granddaughter Helen. Courtesy of Paul Picton.

A range of dwarf varieties were developed by H. Victor Vokes of the War Graves Commission for use in military cemeteries in the early 1920s, derived largely from *Symphyotrichum novi-belgii* and dwarf forms of *S. dumosum*. The discovery of *S. novae-angliae* 'Purple Dome', found by the side of a road and introduced into cultivation in the late 1980s, has further increased the potential for compact varieties. German breeders worked with several species to produce dwarf varieties for cemetery use as well.

Eurybia divaricata, one of several species asters typically found in shaded habitats. Appalachians, North Carolina.

The term "small-flowered asters" is sometimes used to describe species that have not been intensively hybridised. As we have seen, a large number of species have been in cultivation for a long time, but the selection of quality varieties from these species, let alone serious breeding, has been erratic, with 1920s breeding of *Symphyotrichum ericoides* (as *Aster ericoides*) in Germany, being the most notable. From the 1990s onwards, however, there has been much more interest in species asters, as they fit into the contemporary naturalistic aesthetic so well. Also worth noting was the interest in *S. cordifolium*, with breeding in Britain and Germany, particularly 'Little Carlow', bred by a Mrs. Thornely in the 1930s in Devizes, England. This was the result of a cross with *S. laeve*, a species which has contributed to *S. novi-belgii* hybrids and some others.

The distinctive horizontal branching of *Symphyotrichum lateriflorum* has made it useful, while a similar habit in *S. oblongifolium* has attracted the interest of designers relatively recently. Some asters, once widespread, have been lost—most surprisingly *S. puniceum*, a widely distributed garden plant (as *Aster puniceus*) in the 18th and 19th centuries. *Eurybia divaricata* and *Doellingeria umbellata* are almost the only asters now encountered as unselected species.

Cultivars of *Aster chinensis* (long since classified as *Callistephus chinensis*) were once hugely popular as summer annuals, with a real decline not setting in until after World War II. Depictions of their flowers, often shown as huge, almost to the point of being bloated, make very frequent appearances in magazines and seed catalogues in the late 19th and early 20th century. They are still grown but are now only of relatively minor importance.

Bamboo Poaceae

Some 1,200 species make up the subfamily Bambusoideae of grasses. Evolving initially in the Cretaceous, bamboos have spread around the world; they are found across the Americas, Africa, and Asia (as far as northern Australia), wherever warm temperatures and high rainfall create the growing conditions on which they depend. The distinctive above-ground growth is built up through the annual production of hollow culms, known popularly as canes; as the plants age and build up more biomass, culms tend to get thicker. Propagation by division, however, sets the plants back, and it may be many years before new culms are as thick as those on the parent stock. Growth is defined as either sympodial (pachymorph species), where culms are tightly packed, emerging from branching root systems, or monopodial (leptomorph species), where a rhizome runs underground, sending up new culms at intervals. The former are mostly tropical (although *Chusquea* is a genus with temperate members), the latter temperate. Montane tropical South America also has climbing bamboos, which send out horizontal branching culms, enclosing trees in a mass of growth. The overwhelming majority of bamboos in cultivation are from eastern Asia; South America, however, has many which may yet find their way into temperate zone gardens.

Bamboos may be prolific and widely distributed, but they are relatively limited in their environmental preferences, as they almost universally avoid drier habitats or climates with consistently freezing winter temperatures. The latter quality is probably nothing to do with hardiness per se, as so many flourish in gardens in cool-winter climate zones; it is more likely to do with the long time scale needed for seedlings to become established, during which time they can easily be out-competed. Bamboos are well known for mass flowering—all individuals of a species, wherever they are in the world, and in whatever conditions, flower at once and then die, leaving vast quantities of seed.

[I]t is not to be wondered at that gardeners should find it difficult to determine the plants. [T]hey arrive in poor condition [and] years may elapse before they develop sufficiently to allow their being recognised. In the meantime, however, they have been named. [O]ften labels become illegible, or detached and assigned to the wrong plant.

(Katayama Nawohito, 19th-century Japanese writer, quoted in Bell 2000)

Bamboos have benefited enormously from deforestation. The hills covered in forests of bamboo, so typical of parts of central China and northern Thailand, are the result of the bamboo understorey taking over after the forest is cleared. Once established, bamboo makes it very difficult for tree seedlings to grow, all but halting the process of succession.

Bamboo has enormous importance in Asian cultures, particularly east Asian, appearing in a vast array of ornamentation and featuring in many traditional festivals as well as folklore and mythology. In China, bamboo is included among the Four Gentlemen (i.e., four seasons) as representing winter, and the Three Friends of Winter (along with pine and plum, which flowers at winter's end). It is seen as a symbol of virtue, combining strength with flexibility, and is a favourite subject for poetry; but it is perhaps most familiar as an object of ink painting, for the skills needed for painting bamboo are very similar to those required for Chinese or Japanese calligraphy. The shoots can, after cooking, be eaten as a vegetable, and frequently are in many Asian cuisines. They can also be further processed by fermenting or pickling.

The range of uses for bamboo is truly incredible, with more being discovered all the time. Its combination of strength-to-weight ratio and flexibility is exploited most dramatically in the construction scaffolding often seen on Asian building sites; it is also extensively used in traditional building construction. Being hard and yet also strong and relatively easy to work, it is used for a huge array of household implements, tools, and weapons; its hollowness has led to many uses in the making of containers and musical instruments. Historically, the material enabled the Chinese to be particularly inventive, for example in making suspension bridges and drills for natural gas, long before anything similar was attempted in Europe. Since the 1990s, the World Bamboo Organization has coordinated research into using bamboo as a sustainable material, either architecturally, as a source of bulk for composite materials, as a substitute for wood, or as a source of pulp for papermaking or textile manufacture.

In the garden, bamboos have had an important role in east Asian traditions since time immemorial. Because of this, they have primarily been used in Western gardens to evoke the Orient and the exotic, although this psychological connection is perhaps finally on its way to being broken.

The early years of cultivation were marked by a great deal of confusion, as noted by at least one Japanese

An ink-on-paper wall painting by Chinese artist Shiao-hsiang, made during the first quarter of the 19th century but in a style which in some ways has changed little for centuries. This is art as an expression of virtuosity rather than illustration or comment. Courtesy of the Rijksmuseum, Amsterdam.

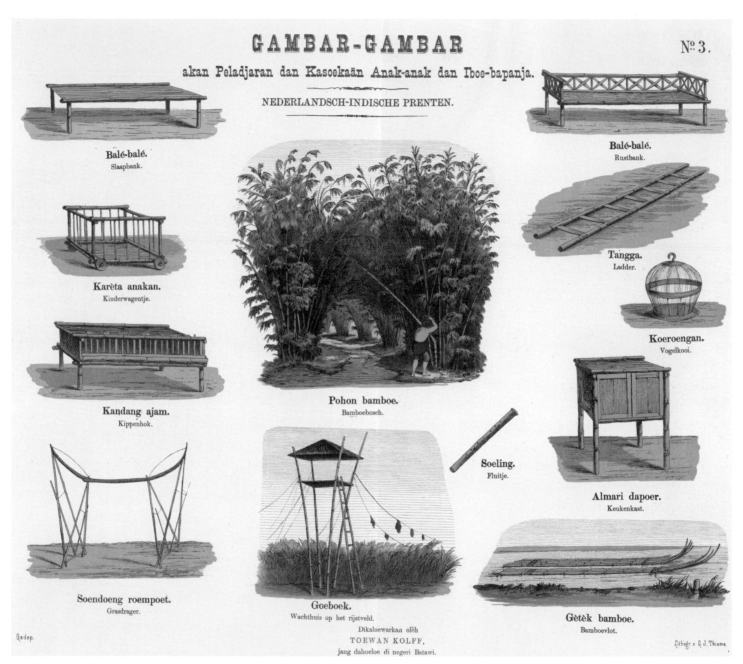

An illustration showing the many uses of bamboo, with a caption in Malay and Dutch, produced in the Netherlands between 1868 and 1881. The country was the colonial power in what was then the Dutch East Indies and is today Indonesia. Courtesy of the Rijksmuseum, Amsterdam.

A Japanese print which gives a modern take on bamboo; by Kamisaka Sekka, of Kyoto, dated 5 May 1909. Courtesy of the Rijksmuseum, Amsterdam.

author (see sidebar), and confusion only deepened as botanists renamed many species towards the end of the 20th century. It is thought that the black-stemmed *Phyllostachys nigra* was the first to be grown in Europe, in 1827. From the 1850s on, a flood of new species hit the United States, Britain, France, and Belgium, with some 230 recorded in France by 1906 (many of these were later lost in cold winters). La Bambouseraie d'Anduze, in southern France, was established in 1856 by botanist Eugène Mazel, who collected every species he could find and set up trials; it was opened to the public in 1902 and is now run as a botanical garden and educational centre. In the United States, late-19th-century introductions were not just for ornamental purposes, as there was interest in growing bamboo as a raw material, e.g., for papermaking, but this did not last long. After World War I, a reaction against exoticism led to a decline in interest in bamboos, especially in Britain, and war with Japan and the closure of China by the communist regime only exacerbated this, with no real revival until the 1980s.

Begonia Begoniaceae

Overwhelmingly tropical (or if not, then found on the edges of the tropics), the distribution of the 1,550 species of *Begonia* is very uneven, with the Americas and Asia (India eastwards) having over 600 species each, but with only 160 species in sub-Saharan Africa and none from Australia. The genus is actually thought to have originated in Africa, and the diversity of American and Southeast Asian species developed from the Eocene onwards. Most are localised endemics with only a few being widespread or adapted to grow in different habitats; of these latter, many are African and deciduous. It is hypothesised that many African species became extinct during arid climate periods, while in Asia and South America, the geological events that helped drive orchid diversity—the development of mountain chains with few links between them—did the same for begonias. Research shows very little gene flow in wild plants, so populations only a few kilometres apart may not interbreed and therefore diversify quickly into new species. When species do meet, hybridisation may begin the process of new species development.

Most species are from moist, shaded habitats: the forest floor, wet cliffs, stream banks. Some are epiphytic or lithophytic. A few are from environments with seasonal aridity. All are clonal perennials, but often grown as annuals in cooler climates.

Begonias are named for Michel Bégon, a French colonial governor, by botanist Charles Plumier (1646–1704), probably to thank him for giving him a post as an official plant collector in the colonies. They have been used medicinally, while one Chinese species, *Begonia fimbristipula*, is commercially available as a herb tea.

Botanical classification is complex and the subject of several recent research projects. These are very much enthusiast plants: currently 66 sections are recognised and some 10,000 cultivars have been raised over time. These and horticulturally important species are grouped into a number of functional groups, as follows: shrub, cane, tuberous, elatior, rhizomatous, and trailing-scandent.

Shrub begonias have upright, branching stems but are not woody. Among them, members of the Semperflorens Group, derived from *Begonia cucullata*, *B. schmidtiana*, and *B. roezlii*, are extremely important as bedding plants, or garden perennials in warmer climates. *Begonia cucullata* was first found in the Berlin Botanic Garden in 1821, as a surprise seedling on orchids imported from Brazil. Ten years later it was being widely cultivated as a hothouse plant. Lemoine crossed it with a number of other species, beginning to sell the results in 1884. Further breeding in Germany, largely by Benary of Erfurt, produced plants that by the end of the century were being widely praised as among the best summer plants for gardens of all kinds and sizes, and suitable for cemetery planting. More species continued to be added to the gene pool through the 20th century.

OPPOSITE *Begonia* 'Gloire de Lorraine'—Victor Lemoine's great triumph with the genus. A book illustration by Louis Fairfax Muckley (1862–1926), an artist and illustrator identified with the Arts and Crafts movement.

Cane types, which produce bamboo-like upright canes, and which may grow up to 4m high with cascades of flower, are most often seen as warm-climate garden plants. While popular, they have never been extensively hybridised.

Tuberous begonias (*Begonia ×tuberhybrida*), derived from *B. boliviensis*, *B. pearcei*, *B. veitchii*, and others, die down to a tuber during the dry season, making them useful for gardeners in temperate climates who wish to overwinter them in a dried condition. The mother species were those found by collectors sent by the Veitches to the cloud forests of the Andes in 1866; spectacular hybrids were then bred from them back home. The plants' high-altitude origin meant that they were tolerant of cool night temperatures,

8-10 inches _Giantflowered_ **DOUBLE BEGONIAS** _for garden, balcony, window-boxes and indoors_								From July until frost starts

← See also page 3

	First size f. garden			Top size for balcony, window-boxes and garden			
	Art. nr.	per 25	per 6	_Art. nr._	per 12	per 3	
DARK BLOODRED	109	15/—	4/—	116	11/—	3/—	
CLEAR YELLOW	110	15/—	4/—	117	11/—	3/—	
INTENSE ORANGE . . .	112	15/—	4/—	119	11/—	3/—	
SNOW WHITE	113	15/—	4/—	120	11/—	3/—	
FIERY SCARLET	114	15/—	4/—	121	11/—	3/—	
PURE SALMON	115	15/—	4/—	122	11/—	3/—	

COLL. 9 First size. 5 each of the colours, mentioned to the left (no white), separately packed **25 for 15/—**

COLL. 10 Top size. 3 each of the colours mentioned to the left (no white) separately packed **15 for 13/9**

N.V.L. STASSEN JUNIOR ◆ HILLEGOM-HOLLAND
Agents for the U. K.: Stassen Limited, Spalding, Lincs. 'The Famous Bulbgrowers'

DOUBLE HANGING BEGONIAS
1 ft. on slopes, 2-2¼ ft. as hanging plants

	per 12	per 3			per 12	per 3
_101_ CLEAR YELLOW	12/—	3/3			From July until frost starts	
_102_ SOFT PINK	12/—	3/3	_104_ INTENSE ORANGE		12/—	3/3
_103_ SNOW WHITE	12/—	3/3	_105_ BLOODRED		12/—	3/3

COLL. 1 3 each of the above-illustrated 5 colours separately packed **15 for 15/—**

Two illustrations from the 1959 catalogue of the Dutch bulb company Stassen, showing double tuberous begonias.

making them very successful as outdoor summer plants in Europe, even as far north as Scotland. The white-flowered South African *B. dregei* was soon involved, widening the colour possibilities. In the late 1870s, Lemoine managed to add yellow and then, in 1876, a double form. It was the double that then really took off, and flower sizes began to reach dinner-plate dimensions—such monsters are still popular with enthusiasts who show them competitively. Even in this golden age of horticultural novelty, they were among the most-talked-about plants. More nurseries tried their hand, and the ensuing frenzy of breeding resulted in a range not just of colour but also of plant and flower shapes.

The elatior types (*Begonia ×hiemalis*) are a cross between tuberous begonias and *B. socotrana* made in the late 19th century. The latter is an extraordinary deciduous species found only on the semi-arid island of Socotra off the coast of Somalia, and the only species found between Ethiopia and India. It was discovered by Isaac Bayley Balfour, regius professor of botany at Glasgow University, in 1880; he was also a pioneer *Rhododendron* taxonomist. *Begonia ×hiemalis* only took off in 1955, when a German nursery developed varieties that were markedly easier to grow, the Rieger series. Tending to flower at cooler temperatures, these winter-flowering begonias, as they are known, are often sold as winter pot plants.

Begonia ×cheimantha is a cross between *B. ×hiemalis* and *B. dregei*, the first of which was 'Gloire de Lorraine' bred by Lemoine. This was one of those hybrids which in itself was not an easy plant, but its mound of prolific pink flowers made such an impact that generations of breeders continued work with it, much of which has been done in Scandinavia. Modern varieties tend to be tetraploid; backcrossing to *B. socotrana* itself has continued to be important, as breeders have found that this creates sterile triploids with a longer flowering period.

Rhizomatous types have densely fibrous root systems and minimal stems, the leaves emerging close to ground level. They are mostly species grown for their ornamental foliage, of which the *Begonia ×rex-cultorum* complex is the best known. Leaf size of these can vary from a few centimetres to nearly a metre in length. *Begonia rex* itself arrived in Europe from Assam in northeastern India in 1856, again apparently as a stowaway on an orchid, although related species were already in cultivation. Rex begonias became popular in the late 19th century as houseplants; presumably their tolerance of shade suited the plants to the gloom of the Victorian parlour. A great deal of hybridisation was done during this period in Belgium and Britain. The smaller, and perhaps somewhat tougher, *B. bowerae*, of Mexican origin, has also been intensively hybridised and grown as a foliage houseplant.

Trailing-scandent types are naturally climbers but are generally grown as trailers in cultivation. Most are derived from *Begonia solananthera*, a species of the Brazilian Mata Atlantica forest region around Rio de Janeiro.

Begonias of all kinds became big business in California after World War I, as they needed little protection in the balmy climate; they were much appreciated locally and could be grown as a crop and sold to the rest of North America. The American Begonia Society (est. 1932) actually started out as the California Begonia Society. Breeding in the state has resulted in new bedding varieties but also larger varieties suitable as garden plants only in warm regions, e.g., the Angel Wing varieties in the cane group.

Southern China and the higher-altitude areas of Southeast Asia harbour a considerable number of frost-hardy species of which only a few are in cultivation, with several having been introduced relatively recently, and for which the potential as garden plants is considerable. *Begonia grandis* is the best known and currently regarded as the hardiest; it has been in cultivation in China and Japan for centuries and occasionally has featured on Chinese ceramics.

Berberis Berberidaceae

The name comes from the Arabic for barberry (*Berberis vulgaris*), a common species of Europe, northwest Africa, and the Middle East. The genus comprises some 450 species of shrubs, half of them found only in China, the remainder across Eurasia, Southeast Asia, East Africa, and the Americas. They hybridise very easily, and it can be very difficult to draw clear boundaries between species. These are among the most primitive of flowering plants. Fossil evidence suggests that they originated in eastern Asia in the late Cretaceous and then migrated from eastern Asia to North America in the Oligocene, via Beringia. They probably arrived in Europe from Asia during the late Oligocene, as proto-Europe and Asia were united at this time. *Berberis* could have migrated to India from eastern Asia, arriving before the last major uplift of the Himalayas in the Pleistocene. *Mahonia* is very closely related and has at times been included in *Berberis*.

Berberis species tend to be found in a similar range of habitats: rocky scrub, woodland edge, and glade. Some, particularly central Asian species, thrive in semi-arid environments. All are long-lived shrubs with considerable powers of continual basal regeneration. While their fruit is welcomed by wild birds, their eating it can spread the plants far and wide, and several species are regarded as invasive aliens in North America. Many species of *Berberis* (and *Mahonia*) are alternate hosts for *Puccinia graminis*, the potentially devastating black stem rust of grain crops, and have been banned in some U.S. wheat-growing states. Public campaigns against berberis plants have been waged on occasion through radio, television, and leafleting.

Berberis vulgaris, scarcely grown now in western Europe, was once widely cultivated as a fruit bush. It is still used for making jam and distilling into spirits in central and eastern Europe and is major element of Persian cuisine, its sour little fruit being characteristically scattered into rice or chicken dishes or made into juice. Some South American species are used similarly. The leaves have also been eaten, and the bark used as a cure for a wide range of diseases and as a source for a yellow dyestuff.

Berberis roots contain berberine, which has antibacterial and antifungal effects that have been understood in China for thousands of years. Recent research indicates this compound has potential as an antidepressant and may be useful in treating diabetes, high cholesterol, and various forms of cancer.

Being dense and thorny, *Berberis vulgaris* was popular as a hedging plant and was used as such by early settlers in North America and New Zealand. Botanical exploration in the late 19th and early 20th century introduced a vast number of new species, some of which have abundantly filled the hedging role for suburban gardens. South America has many attractive hardy species, including *B. darwinii*, discovered by Darwin himself. *Berberis* ×*stenophylla*, a hybrid between it and another Chilean species, turned up spontaneously in 1860 in a Sheffield nursery and is now widely grown. The most valuable introduction to the nursery trade has been the deciduous *B. thunbergii*, identified in Japan by Thunberg in 1784 but not introduced until a century later. Its tendency to throw interesting mutations has been a boon for gardeners, with cultivars selected for different leaf colours and habits (especially dwarf).

> A brilliant family of shrubs so numerous that perhaps no one garden can show a half of their beauty.
>
> (Robinson 1921)

BERBERIS DARWINI *Hook*.

⚥ *Chili (Chiloë)*.   *Plein air (Nord)*.

VII, 663.

Berberis darwinii, named for Charles Darwin, who was the first European to see it, in 1835. It is found in Chile and Argentina and is one of the most widely grown South American shrubs in temperate zone cultivation. Illustration from van Houtte's *Flore des serres et des jardins de l'Europe* (1860).

Bergenia Saxifragaceae

Originally classified by Linnaeus as *Saxifraga*, this group of evergreen perennials were hived off into *Bergenia* by German botanist Konrad Moench in 1794, although this was not recognised by the British, who for a while classified them as *Megasea*, when they were not still calling them saxifrages. The name commemorates 18th-century German botanist Karl August von Bergen. The 10 species are found across eastern Asia, from Afghanistan up to Siberia. They tend to be found as woodland plants, generally at altitude towards the southern part of their range, and appear to have a particular affinity for rocky places—the almost woody rhizomes and their habit of growing above ground is well suited for finding suitable places to root on steep slopes. All are clonal perennials, with a combined competitive and stress-tolerant character, with rhizomes which may be constantly on the move but which maintain highly persistent clumps.

Plants are used as a tea substitute by the Dzungarian people of central Asia; the leaves are full of tannin (research in wartime Germany suggested that their cultivation for this purpose would be economic for the leather industry). Bergenia is used in Tibetan and other regional herbal medical systems as an anti-inflammatory agent; widely collected, bergenias may face local extinctions as a result.

German botanist Johann Georg Gmelin (1709–1755) is credited with introducing the first, *Bergenia crassifolia*, into cultivation, from his travels in Siberia. The 19th century saw more introductions, from China and the Himalayas, with active hybridisation and selection of cultivars for garden use but also as winter pot plants for cool conservatories beginning during the 1890s in England and Germany. Northern Irish nurseryman T. Smith bred many in the early 1900s, although it is not clear how many have survived. The Arends nursery in Germany bred several of the most successful crosses during the latter part of the 20th century.

OPPOSITE A bergenia cultivar and *Trillium grandiflorum* from *Beautiful Garden Flowers for Town and Country* (Weathers 1904), illustration by John Allen. This would have been a notably early example of a book with colour illustrations throughout, using the new, if rather crude, process of chromolithography.

Buddleja Scrophulariaceae

Named for Adam Buddle, a 17th-century English cleric and amateur botanist of great repute. Various spellings have been used: buddleia, buddleya, even buddlea. This is an old genus, dating back to the late Cretaceous, and consequently with a very wide distribution; the 100-odd species are mostly east Asian with some in montane areas of Africa and the Americas. All are woody, mostly shrubs, with a few trees. All are long-lived and regenerate well from the base, which in the case of the firmly shrubby species, means that they can be effectively treated as herbaceous, given that they flower on current year's growth. Preferred habitats tend to be higher-rainfall areas. They are plants of scrub and woodland edges, and in the case of *Buddleja davidii*, and possibly others, of limestone gorges. The fine seed and ability of these plants to root practically anywhere indicate a strong pioneer tendency.

The Asian species have flowers mostly in pale colours, scented to attract pollinating insects, whereas American ones are more likely to be unscented and in the yellow to red spectrum—attractive to hummingbirds, and with longer tubes, too. All the species in cultivation are popular with butterflies; they are among the few shrubs that flower at a time when butterflies are at their peak in northern temperate gardens, and it doesn't hurt that their small, shallow flowers are of the type butterflies favour. Despite this, *Buddleja davidii* is unpopular with some ecologists because of its supposed invasive tendencies, although in reality, habitats where it is liable to spread and effectively compete are few. A pioneer of environments with limited soil, it tends to be out-competed by other shrubs in "normal" growing conditions.

Evergreen *Buddleja coriacea* and several related species are found up to 4,500m in the Andes of Peru and Bolivia, as remnants of the forest which would have covered much of the region prior to the expansion of agriculture by Pre-Columbian civilisations; it is a traditional source of small-scale but high-quality timber in the Andes. *Buddleja officinalis* has various uses in traditional Chinese medicine.

Buddleja globosa was introduced to Britain from southern Chile in 1774. There was then a long gap until Asian species began to be introduced at the end of the 19th century. The most familiar species is *B. davidii*, found by Augustine Henry in the late 1880s in southern China; buddleia, as it is universally known, got its big break with World War II, as it found the ruins and piles of rubble left by bombing very much to its liking. In England, it became almost a symbol of the postwar recovery period.

A number of hybrids have been developed between various species; many forms of *Buddleja davidii* have also been selected, largely on the basis of flower colour—this is naturally a very variable species anyway. The yellow *B. ×weyeriana* hybrids are the result of crossing it with *B. globosa*; the initial cross was made by British Army officer Major William van de Weyer, on leave during World War I. Crossing with Chinese *B. lindleyana* has increased the pink-purple colour range. Given the long summer flowering period, there is considerable commercial interest in buddleia breeding, much of it aimed at achieving sterile plants, as some U.S. states have banned sales of conventional varieties because of worries over invasiveness. Another goal is small-growing varieties, such as the Nanho series. Radiation and chemically induced mutations have proved useful ways of generating new plants.

Buxus Buxaceae

Buxus (from the Latin for the plant) is essentially a tropical genus of 30 species, of which only a minority are hardy, found across Eurasia, Africa (including Madagascar), and the Caribbean down to the northern part of South America. Cuba is a centre of diversity. The temperate species are evergreen shrubs or small trees of the woodland understorey, stress-tolerant and slow-growing.

Since box is evergreen, it has frequently been used as a substitute for other foliage, which, usually for reasons of climate, happens to be unavailable, e.g., as a substitute for palm fronds on Palm Sunday, or to replace rosemary for throwing into a grave. The wood is so hard and dense that it will sink in water; being fine-grained and lacking growth rings, it has been used for small precious objects, including musical instruments and boxes as well as for handles and heads for specialist tools. It is regarded as the very finest wood for making engraving blocks, which is its main use these days. There have been minor uses in herbalism.

The first topiary in Europe was performed by the Romans, most usually on *Buxus sempervirens*. The dwarf 'Suffruticosa', particularly valued for parterres, was known by the 13th century at least, and it was vital to the formal look that was so fashionable in gardens of the early modern period, the Renaissance in particular. By the 17th century topiary was very firmly in vogue, as a way of shaping box for hedging, for parterres, and for sculpture. Some eight varieties, selected for variegation or leaf shape, were known by the end of the next century. The plant evoked a range of reactions from early writers; many disliked its smell, and Stephen Switzer (1718), for one, found Hampton Court "stuffed too thick with box." Although the clipped style was dictated by the wealthy, anyone who had the space for a garden copied it, and after such formality began to decline in the 18th century, its popular use in graveyards in parts of Europe (e.g., Germany and Denmark) survived.

Boxwood (as it is known in North America) was very popular in the early garden styles of the United States, especially in the South during the 19th century. Its popularity declined during the 1900s until the last two decades of the century, which saw a major revival, linked to a strong interest in garden history. In Europe it had a major revival in the 1920s Arts and Crafts garden style, and then another decline until the 1980s, when, as in the United States, an interest in historical styles led to another period of interest, as witnessed by the founding of the European Boxwood and Topiary Society in 1996. Notable has been a great increase in variety selection, choosing cultivars primarily on the basis of plant form (which can be very varied) and leaf size. Enthusiasm has been somewhat dampened by the appearance of a destructive fungus (box blight, *Cylindrocladium buxicola*) in Britain in the 1990s; it has since spread to the rest of Europe and North America.

A boxwood parterre, in classic European style, a design strategy which has proved remarkably successful for centuries, on the Hampton estate, in Towson, Maryland, designed by owner Charles Carnan Ridgely, a governor of the state. The photograph was taken c. 1915 by Frances Benjamin Johnston (1864–1952), one of the first female professional photographers in the United States. Courtesy of the Library of Congress.

Calendula Asteraceae

This English marigold, as it is sometimes known, is the "original" marigold (as opposed to *Tagetes*). The genus includes around 20 species of annuals and short-lived perennials, all herbaceous, stretching from the Middle East, around the Mediterranean, and across to the Atlantic Isles. The name is from the Latin, *calendae* ("the months"), as it is possible for plants to be in flower every month of the year, in a mild winter climate at any rate. *Calendula officinalis* is a classic Mediterranean winter annual, germinating in autumn and flowering in early summer; in maritime climates it can stagger on until past a second winter solstice. It seeds and overwinters readily in cool-winter climates. Despite being much buzzed around by bees, plants are only a modestly good nectar and pollen source.

Traditionally this was the plant with which the gods, and sometimes their devotees, would be garlanded in India, but the gods are a fickle bunch, and these days they prefer *Tagetes* marigolds—the temple suppliers find them a more productive crop. The flower appears in a range of superstitious practices across Europe but without any particular pattern. The plant has been much used in herbalism down the ages and is still widely promoted for improving the complexion and in the treatment of wounds; there is evidence that calendula-based products can help relieve some long-term skin conditions. And as Gerard's quotation tells us (see sidebar), calendula marigolds were major culinary players, decorative rather than nutritional, a use which has been revived in recent years.

Plants were definitely known in German cultivation from the year 900; they were probably in British too. They became one of the most widely grown garden ornamentals of the early modern period, with yellow and orange forms and doubles known by the end of the 16th century. Serious breeding began in the mid-19th to perfect and stabilise doubles, and to increase the range of colour and flower size. The popularity of the plant, particularly vis a vis its "rival" the *Tagetes* marigolds, tends to reflect the interest in

> The yellow leaves of the flowers are dried and kept throughout Dutchland against winter, to put into broths and for divers other purposes, in such quantity that the stores of some grocers or spice-sellers contain barrels filled with them. [N]o broths are well made without dried Marigolds.
>
> (John Gerard, *The Herball, or Generall Historie of Plantes*, 1597)

old-fashioned, romantic- style gardening, this being very much a plant associated with simple, unpretentious cottage gardens, and one which has not proved amenable to the compaction of form which *Tagetes* has shown.

Camellia Theaceae

Camellia contains 125 species: about 104 in China, with more than 80% concentrated south of the Yangtze; some in Korea and Japan; and then more in Southeast Asia. The genus is thought to have originated in southern China in the Cretaceous. In the name, Linnaeus commemorates 17th-century Czech botanist Georg Kamel. Now with more than 3,000 cultivars, camellias are classified on the basis of flower shape: single, semi-double, double, paeony form (with a mass of irregular petals and petaloids), anemone form (one or more rows of outer petals, with inner mixed petaloids and stamens), rose form (overlapping petals with stamens visible), and formal double (rows of overlapping petals with hidden stamens). These flower shapes are often found in more than one of the species-based gene pools. The current range of species in cultivation encompasses three main species, each with a long history of east Asian cultivation: *Camellia japonica* (2,000 cultivars), *C. reticulata* (400 cultivars), and *C. sasanqua* (300 cultivars). There are also various miscellaneous hybrids and species, most of which are

Camellias at a nursery in Vauxhall, London, as shown in the *London Illustrated News* of 1860. The print illustrates the sheer scale of production and shows just what a popular plant this was at the time. The vast majority of plants sold were probably kept only for a few years, and many might well have been disposed of after flowering, much as cyclamen and pot azaleas are treated today.

single-flowered. Unlike roses, rhododendrons, and many other long-cultivated genera, there is as yet not that much interspecific breeding (apart from the notable *C. ×williamsii*), so we cannot talk of a general camellia gene pool (yet).

Camellias are typical evergreen understorey trees, being an important part of the forest community, usually in regions with moderate to high year-round rainfall. They are relatively slow-growing, appear late in the successional process, and can live for centuries. They show limited basal regeneration, so they are not technically shrubs, although often grown as such. They do, however, resprout if cut, which allows them some chance to survive the current onslaught of deforestation in their homelands.

Tea is a *Camellia* species—*C. sinensis*. It is open to question whether the other species could serve up a decent cuppa if anyone tried, Shirley Hibberd (1879) referring to them as "possible tea-plants that nobody wants." Tea drinking, known since the 6th century BC, has spawned a vast and complex culture in the Far East. Drinking tea has been credited with a wide range of health benefits, but nothing evidence-based of great significance. The oil that can be expressed from the seed, however, particularly from *C. oleifera*, is now regarded as the healthiest of all edible oils; it is also used in cosmetics and for specialist lubrication applications.

Literary and pictorial evidence suggests that camellias have been cultivated in China as ornamentals for at least 1,800 years. The Song dynasty saw a lot of creative breeding, probably mostly with *Camellia japonica*, and the use of grafting to produce plants that combined multiple varieties; the Song capital of Hangzhou became a centre for growing and trading the flowers. *Camellia sasanqua* was also cultivated during this period. The next major stable dynasty, the Ming, saw the first books published on camellias.

In Japan, *Camellia japonica* was taken up by the samurai in the 12th century and was much further developed in the Edo period. The Higo clan were very interested in them, and it was they who introduced *C. sasanqua* into cultivation in the 17th century. The second Edo shogun, Hidetada Tokugawa, was particularly fond of camellias, and the *daimyo* (lords) would try to curry favour with him by bringing him unusual varieties. A monk in Kyoto, Anrakuan Sakuden, published a collection of essays on camellias in 1630; around 100 varieties were known at this time. By the end of the Edo there were around 700, including many with flecked or blotched petals and many doubles. Meanwhile, the much-persecuted Christian community used camellias to symbolise resurrection and perseverance.

Camellias, seen as herbarium specimens or in paintings, tempted Western gardeners for some time before they could be obtained. It was for them as much as for anything else that European plant hunters spent years waiting, as virtual prisoners in the tiny areas of the concession ports allowed them by the Chinese and Japanese rulers. The first to flower in England was in 1739 at the home of James, Lord Petre, in Essex, although the species is not known; it was almost certainly raised from seed of Chinese origin. His gardener, James Gordon, went on to establish a nursery after his employer's death, introducing the plant to commerce. Many early plants came to England on the East Indiamen (the ships owned by the East India Company, which imported tea), and early records always name the captains, for it was on their mercies that the plants were thrown during the many months they spent at sea, on their orders that plants were given precious fresh water, or, in a typhoon, thrown overboard.

> The ups and downs of camellia cultivation were also affected by the number of times of riots. When the peasants revolted frequently, people's interest in the camellia cultivation subsided.
>
> (Kashioka and Ogisu 1997)

The camellia was at the peak of its popularity during the 1840s, although only as a greenhouse plant. Nurseries in London and Paris had huge plants, from which they harvested cutting material (fortunately easy to propagate) and buds to sell as boutonnieres. They were in fact immensely popular all over Europe. Above all the flower symbolised luxury, even decadence, as illustrated by the French novel *The Lady of the Camellias* (1848) by Alexandre Dumas.

German explorers of the East also brought back camellia seed; the first flower recorded there was in 1792. There is also the (possibly apocryphal) story of Jacob Friedrich Seidel (1789–1860), who trained as a gardener in Paris under the Napoleonic regime, and in 1812, following the declaration of war between his native Saxony and France, fled home with camellia cuttings in his knapsack, stuffed into potatoes to keep them moist. Seidel in any case became a pioneer of glasshouse cultivation in Germany, with camellias a particular focus. Using propagation and overwintering methods apparently gleaned from Japanese sources, he went on to mass-produce them, in particular exporting vast numbers to the Russian aristocracy. The oldest camellia in Europe is at Pillnitz Castle near Dresden, planted in 1801; it has suffered various vicissitudes but is currently thriving in a purpose-built glasshouse.

Surprisingly, it took a while for people to realise that the camellia is actually hardy, and their cultivation outside in northwestern Europe did not really take off until a long time after the camellia boom had peaked—in Britain at least, they had almost disappeared from nursery catalogues by the 1890s. Most of what was grown had been cultivars of *Camellia japonica*; *C. reticulata* (introduced to Britain in 1820) was more difficult and certainly did benefit from being grown under glass. *Camellia sasanqua* was known but not popular until the 20th century.

The great breakthrough for amateur gardeners came in Cornwall, where J. C. Williams of Caerhays Castle began making crosses in the 1920s. Crossing the recently introduced *Camellia saluenensis* with *C. japonica* he created a series of hybrids, since named in his honour, *C. ×williamsii*. Combining hardiness with vigour, floriferousness, and the tidy habit of dropping their dead flowers (those of *C. japonica* rot away on the branch), these newcomers finally democratised the plant.

The first camellias arrived in the United States in the final years of the 18th century, with New York becoming the main centre for their cultivation—in glasshouses. Introduction further south and on the West Coast brought the camellia to lands where it flourished well as a garden plant. Among other camellia-growing nations, the North Island of New Zealand needs to be recognised as an important centre; old specimens can be seen in churchyards and even around abandoned homesteads.

Breeding over the last few decades has concentrated on hardiness, tolerance of sustained humidity, and yellow flowers. Introductions from the wild, from northern Japan and Korea, have brought in genes from hardy plants, while several other, often rare, Chinese species are also being experimented with; most of this work is being done in the United States.

Campanula Campanulaceae

Campanula has around 500 species, all herbaceous. The name is from the Latin for "little bell." Blue is the overwhelmingly dominant flower colour of the bellflowers, owing largely to the anthocyanin violdelphin, which also colours the unrelated *Aconitum* and *Delphinium*. Although found across the northern hemisphere, they are very much plants of cooler and montane habitats. A very clear centre of diversity ranges from the Balkans to the Caucasus; there are only around 10 North American species and 20 Chinese. The very closely related *Adenophora* tends to replace *Campanula* in Siberian boreal forest and some other Asian habitats. Indeed, some species of *Adenophora* (and *Wahlenbergia*) were once included in *Campanula*, while its only subshrub, the oddly plastic-looking *Azorina vidalii* from the Azores, now has its own genus.

Most are clonal perennials; some are biennials or annuals; and there are some monocarpic and short-lived perennials. The level of clonality varies enormously: *Campanula rapunculoides* (now rarely seen) was one of the most famously spreading plants of the Victorian-era garden, while *C. portenschlagiana* and, even more so, *C. poscharskyana* are famous for popping up through cracks in walls and paving—the latter apparently makes good food for pet rabbits. Many alpine species spread strongly, their questing roots generating shoots readily when exposed to light.

Campanulas often have fail-safe pollination: they are self-fertilising if a pollinator does not visit to ensure seed set. They are not noted as particularly good nectar or pollen sources and are of most value to wild bees in difficult habitats. The larger species are plants of open woodland, woodland edge, and tall-herb flora–with a competitive or competitive-pioneer character. Grassland species tend to be smaller. The vast number of species in the southeastern European centre of diversity are overwhelmingly alpine, with a definite preference for limestone; many are difficult in cultivation; a good number are monocarpic. The Asian adenophoras are campanulas by any other name to the gardener, all remarkably alike, and often with a running root—plants of woodland edge or grassland habitat.

Campanula rapunculus (rampion) was grown as a root vegetable across Europe and regarded by the doyenne of Victorian cookery writers, Mrs. Beeton, as excellent. The Brothers Grimm recorded a fairy tale of a pregnant woman who developed such a craving for the root that her husband had to steal some from neighbours' gardens, one of whom turned out to be a witch; on being caught he had to promise her the baby, a girl, who was named Rapunzel after the plant. It is now all but unknown as a vegetable.

A campanula, rendered as an abstract by Theo Colenbrander (1841–1930), as a design for ceramic manufacture. Colenbrander is regarded as the first industrial designer in the Netherlands. Watercolour on paper, dated 31 January 1919. Courtesy of the Rijksmuseum, Amsterdam.

The annual or biennial *Campanula medium* was probably the most widely grown campanula until the early 20th century and a classic flower of the European cottage garden tradition. An early (c. 1860–70) photograph by Charles Aubry (1811–1877), a pioneer in the field who worked as a designer for fabric, carpet, and wallpaper manufacturers in France. Courtesy of the Rijksmuseum, Amsterdam.

Campanula medium (Canterbury bells) has a long history in cultivation, from the late Medieval period at least. The common name is a reference to the small bells that pilgrims to Canterbury used on their horses; the German translates as "Mary's bellflower." Doubles appeared in the 17th century. *Campanula persicifolia* was introduced in the 16th century, from dry woodland edge habitats in Russia; Arends and Lemoine both bred it to produce a modest range of hybrids, many of which are now presumed lost. Short-lived *C. pyramidalis* (chimney bellflower), now rarely grown, was from the 17th to the 19th century used in parterres and as a temporary houseplant, often for the fireless grate in the summer (hence the common name).

Campanula carpatica, found in the Tatras Mountains of central Europe, was introduced in the late 18th century, its compact habit and long flowering season exploited during the next century as a bedding plant; Arends and Foerster were among many who worked on it. It received a lot of attention as it was small enough to be a pot plant but big enough to be useful outside too, fast-growing enough to use for bedding but long-lived enough to be a good rock garden or front of border plant. A great many varieties were bred, but most have now been lost.

A major flurry of introductions from southeastern Europe and the Caucasus marked the latter part of the 19th century, mostly alpine species needing careful cultivation; there has undoubtedly been a huge loss of alpine species from cultivation since.

An illustration from *The English Flower Garden* (Robinson 1921) showing the traditional use of *Campanula pyramidalis* as a temporary indoor plant.

Canna Cannaceae

The name comes from the Greek for "reed." Around 20 species are currently recognised, all from the warmer parts of the Americas, from South Carolina to northern Argentina. All are herbaceous perennials. Cannas are competitive wetland plants that are adapted to climates and habitats with extreme seasonal variations in moisture. Most typically die down to underground tubers in dry conditions, growing extremely rapidly when the soil moisture content rises again. This speed of growth was recognised early on by Europeans and led to the plant's popularity for bedding out—they can be grown surprisingly far north. Commercially, cannas are grown for their tubers—for industrial starch production, as animal food, and sometimes as human food, e.g., in making gluten-free pasta. The shoots may also be eaten. Mature foliage is fibrous enough to be used as a raw material in papermaking. Flowers are pollinated by insects, birds, and bats.

OPPOSITE *Canna* 'Picasso' from Kellogg's catalogue—the "Garden Beauty Book" of spring 1946. The nursery was based in Three Rivers, Michigan.

Canna indica (Indian shot) can become invasive in warm regions where it has been introduced. Its immensely hard seeds have found a variety of uses, including for jewellery, in musical instruments, and as shot (hence the common name). More recently, the plant has been used in phytoremediation systems to remove nutrients and contaminants from waste water in industrial facilities. It

Cannas from the 1959 Stassen catalogue, showing a variety of both floral and foliar colour.

first appeared in Italy in the mid-16th century, probably via contacts with the Spanish Empire in the Americas. By the end of the 18th century, German writers were promoting it as a summer bedding plant whose dormant tubers could be lifted and stored inside over the winter. The first serious breeding appears to have begun with Théodore Année, French consul to Chile, who introduced several species to France and went to work on them in his retirement in the 1850s. The first hybrids were tall (to 3m) and grown primarily for their coloured foliage. The Crozy nursery in Lyon continued his work, starting about 1870, but using additional species and producing good flowers and some dwarfs (around 1m high). Other French, German, Italian, and British breeders continued to work on the plant.

By 1864 there were 19 species and an unknown number of hybrids in cultivation in Europe. The plants were particularly popular with public garden directors in Germany and France during this period, the nurseries near La Porte de Muette in Paris producing tens of thousands for planting out in parks throughout the city. Cannas were immensely useful for giving an exotic look to gardens, becoming very popular for this purpose in

Britain from the 1880s; Robinson devoted 23 pages to them in the 1879 edition of *The Subtropical Garden*. They also became fashionable in many of the European colonies.

In 1893 cannas made a big splash at the World's Columbian Exposition in Chicago, where there was a bed of them 300m long. The 20th century started with something around 1,100 recorded varieties. Over the next few decades, Antoine Wintzer, a breeder in Pennsylvania, produced around 100 new hybrids—his 'Wyoming' is still widely grown today. There are now thought to be approximately 2,500 canna cultivars, but the number of synonyms makes classification and ennumeration difficult.

> "The Canna is the coming flower," thus we wrote in 1893. We have since been obliged to amend our assertion. The Canna has *arrived*; not only that—it has come to stay!
>
> (*Kelway's Manual*, 1898)

Ceanothus Rhamnaceae

Around 55 species of trees and shrubs are found from southeastern Canada down to Guatemala, with a very distinct centre of diversity in California, where there are 41 species. The very wide range of climatic conditions in the region is reflected in many localised or specialist distributions, e.g., those found only on serpentine or other ultramafic rocks. It is thought that *Ceanothus*, named after a plant mentioned by Theophrastus, underwent diversification during the Pliocene as western North America dried out. Hybrids tend to be fertile, so many species may be the result of hybridisation-driven evolution. In California some species have evolved a range of geographical forms to exploit a number of habitats, including mat-forming ones at the coast, which can clone themselves by rooting from the branches. Plants are found in wooded habitats: subtropical forests in Mexico, open woodland in eastern North America, and of course chaparral.

Species in cultivation tend to be those from the western part of the range, where the climate is distinctly Mediterranean—the shrubby chaparral habitat of California and Mexico. Like many other Mediterranean shrubs and subshrubs, they have relatively short lifespans and a tendency to die suddenly in temperate zone gardens; in the wild, however, it is known that they survive for at least 90 years (albeit lifespan does vary between species). It is possible that different factors affect their longevity in cultivation and that the sudden deaths which often occur are caused by an above-optimum availability of water and nutrients, causing premature deterioration.

Ceanothus are metabolically very efficient, able to photosynthesise at higher levels of drought stress than most other species in the chaparral; they can even photosynthesise right through the summer, making the productivity of their habitat greater than comparable habitats elsewhere in the world. They also have nitrogen-fixing bacteria on their roots. Many are able to regenerate from the base after fire, particularly those with lignotubers (woody, nutrient-storing roots). The flowers attract a wide range of pollinators, while Native Americans used them as a source of detergent.

The first Californian ceanothus to be introduced was *Ceanothus thyrsiflorus*, by Johann Friedrich von Eschscholtz (of California poppy fame). It is still regarded as one of the best. William Lobb, collecting for the Veitches between 1849 and 1863, made some extraordinarily good selections, spotting natural hybrids and exceptional forms: *C. ×veitchianus, C. dentatus* var. *floribundus* (both of these never seen again in the wild), and *C. ×lobbianus*.

Back in Europe, growers in Britain and France went to work and created a succession of hybrids that combined good looks with hardiness; early-19th-century French and Belgian breeders worked on *C. coeruleus*, *C. americanus*, and *C. herbaceous*—'Gloire de Versailles' being one successful result. In California, English horticulturist Theodore Payne (1872–1963) was the first to grow and sell native selections, followed by other growers who started trials in the 1920s. During the 1960s and 1970s several California botanical gardens started to launch hybrids, and since then nurseries have continued to make selections, often for the growing field of xeriscaping.

Chrysanthemum Asteraceae

One of the world's most successful commercial flowers, there seems justice in the plant's name being derived from the Greek for "golden flower." Once larger, *Chrysanthemum* is now a much-reduced genus, with around 30 familiar herbaceous or subshrubby species recognised from eastern Europe across to the Far East. Polyploidy and hybridisation are common, so the origin and classification of this and related genera is still in flux. *Chrysanthemum* species are generally long-lived and often clump-forming. Some have persistent semi-woody growth, but others tend to die out in patches. Generally they are from woodland edge habitats, although several are common along seashores in Japan. Species are found in a number of climate zones, with many of those which have

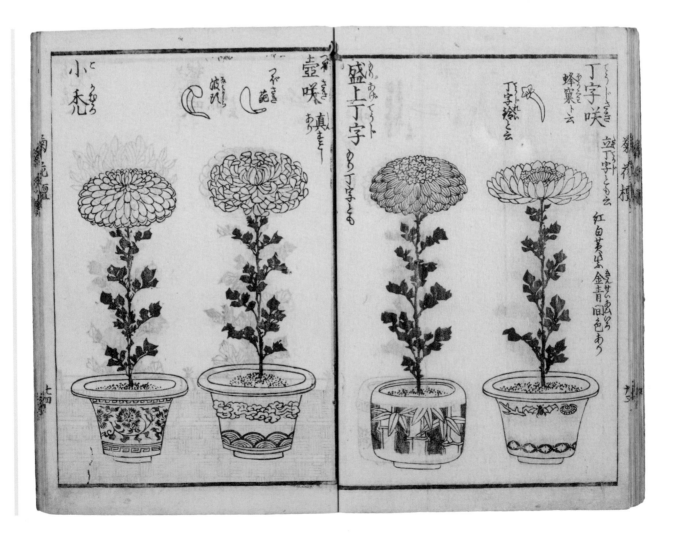

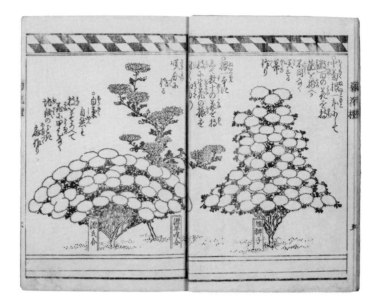

A chrysanthemum show and various methods of growing the plants, from the book *Cultivation of Chrysanthemums*, published in 1846. Courtesy of Chiba University Library.

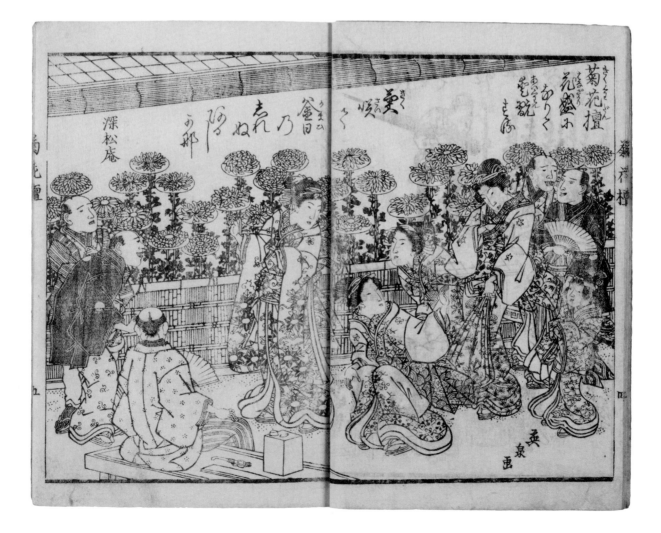

contributed to the cultivated gene pool from the Far Eastern humid subtropical zone. The complexity of the natural genetics is echoed in that of cultivated plants, and the number of cultivars is very unclear, particularly the commercially important derivatives of *C. indicum*, the key species; the fact that it is not reliably hardy has prompted crossing with related species.

Chrysanthemum flower tea has long been popular in the Far East as a mild stimulant, or to soothe sore throats; in traditional Chinese medicine it is regarded as especially beneficial for the liver and eyes. Some *Chrysanthemum* species have been shown to possess antibacterial, antifungal, and antioxidant properties, while *C. indicum*'s use as an anti-inflammatory in traditional Chinese medicine is supported by research. That chrysanthemums have an interesting chemistry is rather suggested by their potent smell; a species formerly classified as one, now *Tanacetum cinerariifolium*, is the source of the insecticide pyrethrum.

Chrysanthemums, rendered almost to abstraction by Nakamura Hochu (fl. 1790–1830), in a print produced in Edo (Tokyo) in 1826. Courtesy of the Rijksmuseum, Amsterdam.

The Japanese emperor Gotoba (1183–1198) particularly liked the flower and started using a chrysanthemum graphic as his own personal symbol. Other emperors followed suit, and in the late 13th century it became the official royal family symbol. Today, in English-speaking countries, the Japanese ruling institution is sometimes referred to as the Chrysanthemum Throne. In the East, chrysanthemums have tended to be symbolic of long life, which is perhaps another reason for the popularity of chrysanthemum tea; in the West, however, they became a funeral flower during the course of the 19th century and so were frequently, superstitiously excluded from the home, even being seen as a curse in Italy.

Cultivation is known to date back to at least the Chinese Shan dynasty (1700–1046 BC). During the Tang, literary sources indicate that ornamental chrysanthemum growing took off, with many distinct varieties being selected. Doubles, a range of colours, and multi-coloured flowers appeared during the Song; there were 400 varieties by 1458, when the first book on the flower was published. Meanwhile, over the water in Japan, the flower became a popular subject for poetry, and a number of customs developed such as the floating of chrysanthemum flowers in sake cups for autumn festivals. Its flowering in the cooler days of autumn made it popular in both cultures and the focus of much seasonal symbolism.

During the Edo period, many new varieties were produced, indeed whole new classes of flower, and novel growing techniques developed. One involved the removal of side shoots, so that one enormous flower develops instead of multiple small ones, a technique later followed by competitive growers in the West. Other techniques aimed at bonsai, or "waterfall" or "bridge" styles, where the plant is trained over a bamboo framework to produce a cascade of flowers. The first Japanese chrysanthemum book appeared in 1717. Classes were defined by flower shape, most probably a reflection of the genetic heritage of the plant. Different classes were also strongly linked with particular regions. Some Japanese historians of the genus have suggested that the different

Chrysanthemums grown in traditional Japanese style at the New York Botanical Gardens. Above is the "waterfall" style, below is the "bridge." These would require pruning and tying in almost daily.

groups originated from up to five native Japanese species. Competitions were held, often in temples, with flowers held in place in a bamboo tube; special display cabinets were also made for them. In the tea ceremony, however, simple, wild type flowers were preferred, in keeping with that tradition's ascetic ethos.

Chrysanthemum growing reached a zenith in Higo culture, when good chrysanthemum cultivation was seen as encouraging four virtues (justice, politeness, wisdom, fidelity), while a man's *kiku* (chrysanthemum) reflected the state of his soul. Even a particular garden style was developed, which specified how the plants were to be laid out and cared for. The most important classes are as follows:

- Saga. Narrow petals form a delicate, loose head. Traditionally cultivated exclusively at the Zen temple of Daikakuji in Kyoto. The plants are descended from a seedling which spontaneously appeared on an island in the temple garden during the reign of Emperor Saga (809–823).
- Edo. Middle-sized flowers, bred first in Edo (now Tokyo). The petals tend to kink, turning back over the centre of the flower, a bit like a punk version of a comb-over.
- Ise. Named for the region where they were first bred and which has the most sacred Shinto shrine; the flowers have traditionally been grown by priests there. The petals are narrow, hang down, and often have frilly or split ends, making for a tassle-like effect. Best appreciated by looking up at them while seated on the floor.
- Osyu. Translated as "great fist." The heavy twisted outer petals often need the support of special wire structures, while the inner ones are tightly curled up.
- Higo. Single, with narrow petals and a simple, unpretentious beauty, although there are intricate rules in growing: heights, numbers of stems, and of flowers per stem.

The first chrysanthemums were grown in the West (in Britain and France) during the 1830s, with the Americans following only a few years later, after several traders brought plants back from China. A key pioneer in Western breeding was the ever-versatile John Salter. Interest grew and was sufficiently strong for the organization in 1846 of a Borough of Hackney Chrysanthemum Society, which later became a national society. This was also the year that Robert Fortune returned from China with additional varieties, including the Chusan daisy, which gave rise to the pompom group; Chusan refers to the coastal island city now known as Zhoushan. The Rubellum varieties originated by chance in 1929 in a garden in Wales; they involved *Chrysanthemum zawadskii*, a Eurasian species, and are the source of some of the best genuinely hardy hybrids, many developed by Perry's Hardy Plants in the 1940s. The hardier Korean varieties, derived from *C. zawadskii* subsp. *coreanum* and *C. japonicum* crossed with *C. indicum*, were launched in the United States in the late 1930s, a reflection of adventurous expansion of the gene pool.

Industrial-scale production of chrysanthemums began in California with the Japanese immigrant Enomoto brothers, who started in the 1880s, first sending flowers out of state in the 1910s. The flower then rapidly took off as a commercial crop, rather eclipsing its growth by amateurs, who in many regions were limited by the plant's lack of hardiness. Today, the flower stands alongside the rose as the most important genus in the global cut flower trade. Cutting-edge and high-tech breeding techniques play an important role in producing new varieties; the use of gamma or x-ray radiation to induce mutations has been particularly successful.

The complex classification of chrysanthemums in the West—based on numbered classes defined by flowering time and flower head form—makes the Japanese system look almost straightforward. In the British application of the system, Section 29, for example, is September-flowering Spray varieties, and 10 is November-flowering

Spiders, Spoons, and Quills. This level of classification is a sure indicator that the flower had become the subject of competitive growing, which was at a peak in the first half of the 20th century, at least in the United Kingdom. In Britain, the growing of chrysanthemums was very much part of a male working-class culture, a continuation of the florist flower tradition, with a strongly competitive element; with the disappearance of this culture following changes in the economy, amateur chrysanthemum growing has gone into a steep decline.

Chrysanthemum indicum varieties, bred in 1930s Germany for hardiness. They were promoted as winter pot plants and for cutting, graveside planting, and garden use. A watercolour by Esther Bartning, *Gartenstauden Bilderbuch* (Garden perennial picture book), by Karl Foerster, 1938. Esther Niedermeier Bartning (1906–1987) illustrated the covers of *Gartenschönheit*, the magazine of which Foerster was editor-in-chief, as well as several of his books. Courtesy of Bettina Jacobi.

Cistus Cistaceae

Cistus (from the Greek for the plant) is a genus of around 20 shrubs and subshrubs, found around the Mediterranean. The plants hybridise ferociously, both in the wild and in the nursery, many "species" being actually natural hybrids. The closely related *Halimium* can also cross naturally with *Cistus* to produce ×*Halimiocistus*. This is a very young genus; an ancestor shared by *Halimium umbellatum* and *Cistus* appeared to have diverged after the Pliocene-Pleistocene boundary, and species diversification then rapidly took off as the proto-Mediterranean region dried out.

Cistus species are pioneers, dominating habitats (such as maquis scrublands) where fires are frequent, often surviving only 15 years. On overgrazed land with degraded soils, certain species (*C. monspeliensis*, *C. salviifolius*, and *C. ladanifer*) may form extensive and dense stands, which protect the soil from erosion, add humus, and provide a cooler microclimate, slowly improving the soil, enabling shade-germinating trees and shrubs to grow and ultimately restore the climax woodland. This lifecycle does mean, however, that they are never long-term garden plants.

The plants have a very complex set of relationships with a wide range of mycorrhizal fungi, which enables them to thrive on very poor soils. Some of the fungi have considerable gastronomic significance, such as truffles and boletes; cistus innoculated with

Cistus ×*purpureus* in the wild, Italy.

An extract from "Still Life with Flowers on a Marble Tabletop." Prominent here is *Cistus ladanifer*, which, at the time this was painted, in what is now the Netherlands, could only have been marginally hardy—this was during the Little Ice Age. Also included are a poppy (*Papaver rhoeas*, looking very much like one of the later Shirley poppies); roses, *Convolvulus tricolor*, peonies, and what looks like an Asian-origin primula (far left). By Rachel Ruysch (1664–1750), a highly regarded painter of flowers and one of the most prominent female artists of the Dutch Golden Age; indeed, during her lifetime her flower paintings reached considerably higher prices than Rembrandt's work. Oil on canvas, dated 1716. Courtesy of the Rijksmuseum, Amsterdam.

truffle fungi are available commercially in France. The highly aromatic gum exuded by several species has been used as a source of incense for millennia and was reckoned to be the nearest vegetable equivalent to ambergis (an aromatic gum derived from whales). It has also been used for treating respiratory complaints. Collecting the gum was problematic, requiring special tools to comb it off the leaves or, alternatively, cutting off the beards of the goats that fed on the bushes and then melting it off the hair.

In cultivation in England since the 16th century, the plants' lack of hardiness has kept them rare. There was a brief burst of enthusiasm for them in the 1820s, as illustrated by a monograph on them published in 1829 featuring over 40 species and hybrids, the plants being painted from those in the Chelsea Physic Garden, which still has a National Collection. The odd selection was made through the 20th century.

Clematis Ranunculaceae

The 320-odd species of *Clematis* (from the Greek for the plant) are overwhelmingly semi-woody climbers but include some herbaceous perennials, herbaceous climbers, and weakly woody small shrubs. Clematis flowers lack petals; what we interpret as petals are the enlarged and colourful sepals. Some have a ring of structures, petaloid staminodes, between the stamens and sepals, creating the effect of a double flower. Some species are notable for their ability to cope with semi-desert conditions and may form an important part of the vegetation in such regions, especially in central Asia. Despite being toxic, some species were used in Native American herbal medicine. The stems were used for making string and rope. The climbing habit of most clematis tends to restrict them to woodland edge habitats, glades, or woodland that is open enough to allow them to grow at ground level and scramble up shrubs or other vegetation. Many have benefited from human intervention creating transition zones, such as along roads. All are long-lived and can be regarded as competitors or stress-tolerant competitors.

The global distribution of *Clematis* suggests it is old—indeed, it is thought to have first evolved in the Eocene—but real species diversification is thought not to have happened until the Miocene. Various quasi-taxonomic divisions have been made, based on an uncertain set of botanical divisions, to try to bring order to this complex genus, whose boundaries have been blurred by much garden hybridisation (some groups, e.g., Atragene and Montana, do not readily hybridise with others, however). The current state of affairs? Horticultural groups based on botanical taxonomy. In each case the group

White-flowered *Clematis songarica* growing with *Perovskia scrophulariifolia* in semi-desert, Kyrgyzstan. *Clematis orientalis* and several other familiar species of cultivation can be found in this region with its extreme continental climate and often severe shortage of water.

is understood as including species belonging to or derived from divisions of the genus. The Large-flowered Division, those clematis essentially derived from *C. lanuginosa* and *C. patens*, is split simply into two groups, by bloom time: Early Large-flowered and Late Large-flowered. The Small-flowered Division is more complicated:

- Armandii Group, subsection *Meyenianae*, mainly *Clematis armandii*.
- Atragene Group, subgenus *Atragene*, e.g., *Clematis alpina*, *C. macropetala*.
- Cirrhosa Group, *Clematis cirrhosa*.
- Flammula Group, section *Flammula*, e.g., *Clematis flammula*, *C. recta*.
- Forsteri Group, section *Novae-zeelandiae*, some of which natives of Australia and New Zealand are non-climbing alpines.
- Heracleifolia Group, subgenus *Tubulosa*, e.g., *Clematis heracleifolia*.
- Integrifolia Group, *Clematis integrifolia* and related species; herbaceous, non-climbing, very sweetly scented.
- Montana Group, section *Montanae*, e.g., *Clematis montana*.
- Tangutica (Orientalis) Group, section *Meclatis*, vigorous Asian species, e.g., *Clematis tangutica*, *C. orientalis*.
- Texensis Group, *Clematis texensis*.
- Viorna Group, section *Viorna*.
- Vitalba Group, section *Clematis*.
- Viticella Group, mainly *Clematis viticella*.

Clematis viticella was in European cultivation by the 16th century, and several colours (blue, red, and purple) were known by the late 18th. Other attractive European species followed into the garden: *C. flammula*, *C. cirrhosa*, and *C. alpina*. But the key introduction for clematis breeding was that of two large-flowered east Asian species: *C. lanuginosa* (a Chinese species introduced by Robert Fortune, with the largest flowers of any wild species—to 15cm across), and *C. patens*; although it is pos-

A watercolour of *Clematis* 'Purpurea Plena Elegans' (the double) and *C.* 'Étoile Rose', both in the Viticella Group, by Graham Stuart Thomas (1909–2003); Thomas was one of the most influential horticulturalists of the 20th century, particularly in his role as gardens advisor to the National Trust, the largest owner of historic gardens in the United Kingdom. Courtesy of RHS Lindley Library.

sible that these are simply forms of the same species. These had been in cultivation in China and Japan for centuries, and it is probable that growers there had done much of the genetic groundwork for later European breeding—as well as possibly creating some genetic bottlenecks. Varieties from this gene pool have a tendency to flower on side shoots (useful) and a vulnerability to the disease clematis wilt (almost disastrous).

Clematis montana arrived from the Himalayas in 1831, brought by Lady Amherst, wife of the retired governor-general of British India. Around this time too, nurseries began to cross available clematis. The most successful cross, indeed one of the most successful clematis of all time, was made in 1858 by George Jackman; it involved an existing hybrid, *C. viticella*, and perhaps crucially, *C. lanuginosa*. Rich purple *C.* ×*jackmanii* is

What are clearly very similar to modern large-flowered clematis are seen in this Japanese book, compiled by Tachibana Tamotsukuni, 1755. Courtesy of Chiba University Library.

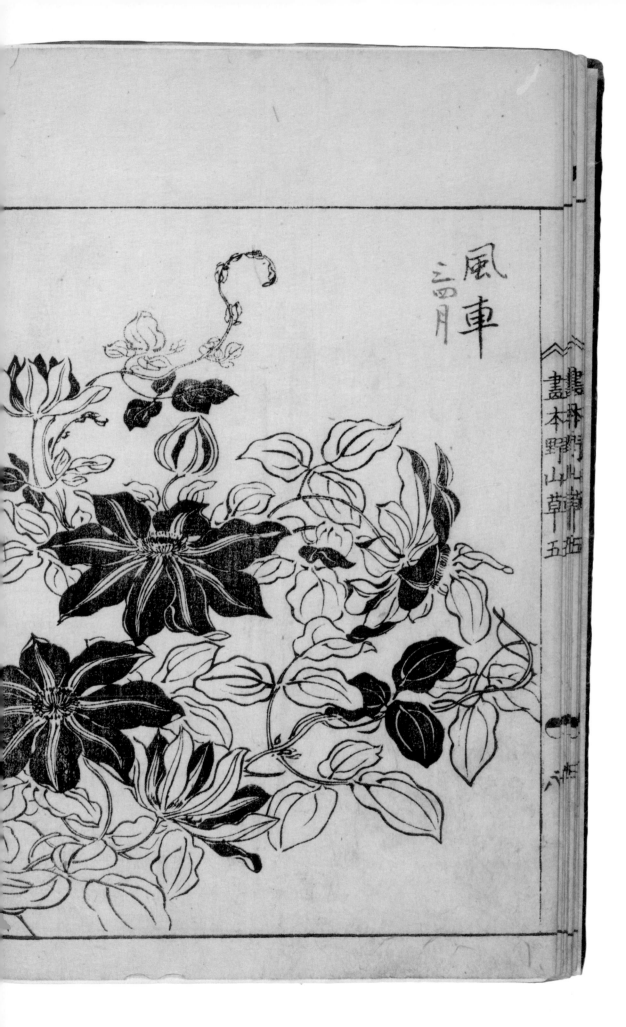

風車
三四月

still in production, despite many hundreds of other hybrids being raised. Jackman went on to write a monograph on *Clematis* in 1872, in which he listed over 200 species and varieties.

English, French, and German breeders worked intensively from 1860 to 1880, producing over 500 large-flowered varieties; many, however, were grafted onto *Clematis vitalba* or *C. viticella* rootstocks, which did not necessarily serve them well, and by 1900 there had been a decline in interest, as lifespan was often short. Most of these early varieties have been lost. There was much less interest in the smaller-flowered species, although Robinson is credited with introducing *C. texensis* from the United States, selecting 'Gravetye Beauty', and promoting the growing of clematis through trees and shrubs. Of the largely central Asian Tangutica Group, *C. orientalis* was brought to Europe by Tournefort, and *C. tangutica* to Kew via St. Petersburg in 1898.

> The most graceful climbing plants of the northern world, for half a century most of them have been lost to our gardens owing to the mistaken mode of increase by grafting these beautiful Chinese and Japanese plants on the common vigorous kind that grows on the chalk-hills of Surrey. Death is inevitable.
>
> (Robinson 1921)

Breeding in the 20th century was less frenetic and more attention was given to smaller-flowered species, particularly the early-flowering Atragene Group and the Viticellas. American breeders, such as Arthur Steffen of New York, have produced many good hybrids. In Europe, Swede Magnus Johnson is regarded as the most important contemporary breeder. Recent developments, such as the work of the British Raymond Evison, have stressed small-growing cultivars for balconies and other confined spaces. The diminutive flowers and mysterious colours of the Viorna Group have also attracted something of a cult following in their North American homeland.

Clematis breeding in Russian-dominated eastern Europe remained almost hidden until the fall of communism in the late 1980s, when much good work in Estonia, Latvia, and Poland became visible. One of the finest varieties of recent years, *Clematis* 'Polish Spirit' (Late Large-flowered Group), was bred by Stefan Franczak, a Polish monk. And back in Japan, perhaps the true home of the most important part of the modern gene pool, clematis are once again the subject of intensive breeding.

Colchicum Colchicaceae

The name of this genus of corm-forming geophytes is after Colchis, a region noted in the Classics, now part of Georgia. The approximately 100 species are scattered across Europe and the Middle East, to the western Himalaya, and in North Africa, down through East Africa to South Africa. The autumn-flowering temperate zone species are commonly known as autumn crocuses owing to a confusion with those members of the true *Crocus* genus, which also flower, without their leaves, at this time. Anyone who gets them muddled, however, will find that in spring the *Colchicum* "crocuses" will produce a mass of broad leaves, quite unlike the demure strips of *Crocus* foliage. There is also widespread confusion over the exact naming of many of the so-called species in cultivation, as many are possibly hybrids.

Most of the species grown are from Mediterranean or Middle Eastern regions, where there is an advantage to flowering in the cool of the autumn. These save foliage production until the warmth of early spring, to take advantage of higher, post-winter soil moisture; post-winter leaf production also reduces snow damage. Those of African origin are more likely to have foliage and flowers at the same time. Habitats include woodland edge, scrub, and grassland.

The compound colchicine is highly toxic but has had a number of uses in herbal medicine, particularly for treating arthritis and gout. It has also been used for deliberate poisonings. One interesting action: colchicine inhibits cell division, which has led to its being used for stimulating the doubling of chromosomes in plant breeding and therefore the creation of polyploids; this has proved useful across a wide range of agriculturally and horticulturally important genera from the 1930s onwards. This effect has also led to interest in it as a potential anti-cancer drug, but research is still in its early stages.

Parkinson had 17 *Colchicum* species in his 17th-century garden; the European native *C. autumnale* was by this time quite common in German gardens as an ornamental. Parkinson and others would have obtained their non-European species through the trade routes that reached down to the Mediterranean or the Ottoman Empire. The end of the 19th century saw new introductions from the Middle East and the Caucasus. E. A. Bowles did much to popularise them in the early 20th century.

Colchicum speciosum in September on the Zigana Pass, south of Trabzon, Turkey.

Conifers

Conifers comprise one of the twelve divisions into which the plant kingdom is categorised by botanists. They are defined as having reproductive structures that are less advanced than the flowers, with seeds borne in cones. For the gardener, conifers are particularly important because of the very large number of dwarf varieties that are grown. In most cases, these are not true dwarfs but merely very slow growing variants, as many gardeners have discovered to their cost. Some of these are the result of genetic mutation and have been propagated from plants that grew slowly from seed; most, however, are the result of somatic mutation, when the plant's production of growth-regulating substances are reduced for some reason, such as injury or disease. This often results in a witch's broom, a congested mass of slow-growing shoots. During the 19th century it was discovered that if cuttings were taken from these growths, the resulting plants would continue to grow very slowly.

Dwarf conifers have the ecological preferences of their normal forms, and since most conifers are plants of exposed and stressful environments, this means that they are particularly useful for those gardening in high-latitude or high-altitude regions, in continental climates with severe winters, or in exposed locations.

The 19th century saw a stream of conifer introductions to European gardens; by mid-century, much new planting was of conifers, at first from North America but then later from eastern Asia. With the discovery that some, notably cypresses, could be easily propagated from cuttings, it became fashionable to use small conifer plants as temporary winter bedding. The opening up of Japan brought with it not only many good conifers, including some dwarfs derived from witch's brooms, particularly of *Chamaecyparis obtusa* (hinoki cypress), but also an awareness of bonsai.

Once it was realised that witch's brooms could be propagated, nurserymen in Europe and the United States began to look out for them in order to find new cultivars to satisfy a growing demand. Dwarf varieties were increasingly available in the latter part of the century, the Lawson nursery in Edinburgh listing around 40 in 1870; there was then a decline but later a revival linked to the interest in rock gardening in the 1920s and 1930s. In central Europe, owing to the harsh winter climate, interest (not just in dwarf cultivars but full-size conifers, too) was particularly strong.

The 1970s saw dwarf conifers reach a height of popularity, partly owing to their promotion by Adrian Bloom, a nurseryman in eastern England, as part of a low-maintenance planting style that involved their being combined with heathers and other low-growing evergreen shrubs. For many the irony was that Bloom's father, Alan Bloom, had been one of the great promoters of perennials from the 1950s onwards. The new planting style was controversial and regarded with distaste by many gardeners, leading to a long-running series of jokes about dwarf conifers being like flared trousers: a fashion mistake that is always threatening to come back.

Convallaria Asparagaceae

The name is from the Latin for "valley." Generally recognised as the only species in its genus, *Convallaria majalis* is found across temperate Eurasia—a classic spring-flowering woodland plant, growing slowly to form very dense mats, displaying a stress-tolerant character. Like many such forest floor plants, it is slow to establish and seeds infrequently but potentially dominant in time. The plant is, however, able to survive in grassland once it loses its forest cover, which illustrates its powers of survival in more competitive conditions.

Although the plant is very poisonous, it was used as a cure for gout. Widely picked in mainland Europe for Whitsuntide celebrations, it appears in a number of Christian myths, e.g., as springing from the tears of Eve, or of Mary at the Crucifixion.

Various colour forms were selected from the 16th century onwards. Commercial production has concentrated on the cut flower and forcing markets. Production of rhizomes for forcing to sell as a flowering pot plant in the cities led to some places in Germany developing specialist growers and dealers; around Frankfurt (Oder), for example, some 25ha were in production in 1914—the scent must have been overpowering in the locality!

An etching of *Convallaria majalis* by Van Eeghen, with printing by Casper Luyken. Illustration from Steven Blankaart's *Dutch Herbal Plants and Spices*, published by Jan Claesz ten Hoorn, Amsterdam, 1698. Courtesy of the Rijksmuseum, Amsterdam.

Cosmos Asteraceae

Twenty-six herbaceous species make up *Cosmos* (which name derives from the Greek for "ornament"), all from the warmer regions of the Americas, from the southern United States down to Paraguay, with a centre of diversity in Mexico. It is thought that Asteraceae first evolved in South America and radiated out to the rest of the world from there. Much of the family has evolved relatively recently; this recent evolution accounts for the limited and clearly defined geographical distribution of many genera, such as *Cosmos*, in the Americas. One species, the dark red chocolate-scented Mexican

Cosmos seed strains from Vaughan's 1936 catalogue, clockwise from upper left: Double Early Flowering Mixed, Orange Flaire, Miniature Golden, Early Flowering Mammoth Mixed.

C. atrosanguineus, is extinct in the wild and exists only as a single clone in cultivation. These are predominantly short-lived pioneer plants, often of seasonally dry habitats.

Some species have uses in herbal medicine, both in South America and in the many countries to which they have been introduced, including for the treatment of malaria. *Cosmos bipinnatus* has been shown to have significant antioxidant activity; its use in skin treatments may possibly be effective in helping to slow signs of ageing.

Cosmos bipinnatus was introduced to Europe in 1789 via the Royal Botanic Garden of Madrid, and through the Spanish Empire to the Philippines, and thence to the rest of Asia. By the mid-19th century it was distributed throughout Europe but was too lax a grower to make a good bedding-out plant; during the 20th it became a popular plant for informal annual plantings, and for large-scale sowings of annuals in both temperate zone and warmer climates. It has naturalised in parts of South Africa.

Cotoneaster Rosaceae

There are around 600 species, both deciduous and evergreen, of this woody plant genus, whose name derives from the Latin for "wild quince" (which is not a cotoneaster). Cotoneasters are found across temperate Eurasia with a centre of diversity in China. They vary from ground-hugging shrubs to 18m-high trees. Some of the former are able to clone themselves by rooting their branches into the ground. They are generally plants of open habitats, with the smaller species being well adapted to growing in exposed rocky locations.

The flowers can attract vast numbers of bees, wasps, and other pollinators, enough to make a noise that can be heard at the other end of the garden. The berries are toxic to humans, but birds love them; particularly useful to both them and gardeners are those species whose fruit ripens only towards the end of the season, when all other fruit sources are eaten. Birds distribute the seeds over a wide area, leading to plants frequently naturalising in habitats where there is little competition. Cotoneasters are increasingly seen as potentially dangerous invasive aliens, in Britain at least, as their adaptability to rocky habitats leads them to smother all in their path.

The limited number of European species attracted little interest from gardeners; the 19th century, however, saw a constant flow of new species from the Himalayas (many via the botanic gardens of Calcutta) and China. The small-leaved evergreen species, such as *Cotoneaster horizontalis*, are quite unlike anything in the European flora and have played important roles in postwar landscaping as space-fillers and ground-coverers; *C. microphyllus* is dubbed "the Architect's Friend" for its habit of hugging walls or any other obstruction in its path, "so densely that nothing can be seen" (Coats 1963). Such uses were not immediately apparent in Victorian England, however, and gardeners did not really know what to do with them at first—Loudon proposed grafting them onto hawthorn standards to make weeping trees!

The 1930s botanical exploration of central Asia by Soviet botanists led to the discovery of more species, but these have been little grown outside the region, and the range of plants in cultivation remains limited, for the most part, to selections of a relatively few species, a reflection perhaps of cotoneasters tending to be thought of as more utilitarian than seriously decorative. The landscape industry continue to use the evergreen species extensively—they are among the plants derided as "green cement" by the industry's critics.

Crocosmia Iridaceae

Native to southern and eastern Africa, from South Africa to Sudan, the eight species of *Crocosmia* can be evergreen or deciduous perennials. The name derives from the Greek for "saffron odour," a reference to the smell of the dried leaves. Crocosmias are plants of forest margins or sometimes even dense forest; however, those that have contributed to the gene pool in cultivation tend to be from the montane grassland-dominated regions of southeastern Africa. They tend to be found on moist soils, such as those along streamsides. All are able to die back during cold or dry seasons to a corm, formed from the base of the stem; they are clonal perennials, with considerable ability to spread over time. *Crocosmia ×crocosmiiflora* is now established as a weed in many temperate climate regions, although rarely a problematic one. The flowers are both insect- and sunbird-pollinated, with some species adapted to pollination by butterfly or bee species with long slender mouthparts. Plants have been put to some uses in African traditional healing, e.g., for treating infertility.

> It is strange that these delightsome perennials are so little appreciated.
>
> (Thomas 1976)

The first European introductions to Britain and Germany were made in the early 1800s. Plants eventually found their way to Nancy, where around 1880 Lemoine crossed *Crocosmia aurea* and *C. pottsii* to produce *C. ×crocosmiiflora*. He then embarked upon what has been described as an incestous breeding programme, but despite his limited range of material, several of his hybrids are still very widely grown, notably 'Solfatare' and 'Gerbe d'Or' (as 'Coleton Fishacre'). With more species arriving during the last decades of the century, others, particularly the leading bulb nursery, Krelage of Holland, carried out further breeding.

George Davison, an English head gardener, worked on the genus in the early 1900s, producing several good hybrids, including 'Star of the East' with flowers up to 10cm across. The 1920s saw several growers enthusiastically hybridising, but interest declined after 1930, until Bloom made some crosses in the 1950s: his 'Lucifer', a hybrid involving the rather muscular-looking *Crocosmia paniculata*, has been enormously successful. Sporadic breeding has continued ever since, but nothing systematic.

Crocus Iridaceae

Crocus (after the Latin for "saffron yellow") is a genus of around 80 species of small perennials, growing annually from corms, native to Europe, North Africa, the Middle East, and across central Asia to western China. Crocuses are spring ephemerals, or in some cases spring ephemerals but with flowers produced earlier—in the autumn. They are found in a variety of habitats, including deciduous forest, scrub, semi-desert, and high-altitude grasslands. Many have adapted well to human-created grasslands; they have little tolerance of competition while in active growth, however, which limits their use in maritime latitudes where grass growth is almost year-round, but not their use alongside perennials whose active period of growth is later. Species from the eastern part of the range are used to hot dry summers and very sharp drainage, tending to be difficult and rare in cultivation.

Saffron, the dried stamens of *Crocus sativus*, is one of the most famous of food additives; once very widely grown, it is now produced only in rural economies where labour is cheap enough for the immensely painstaking task of plucking the stamens from the flowers. Famed as a cure-all in Medieval times, when those who adulterated or forged it could be burnt at the stake, saffron has always been expensive. The origin of *C. sativus* is something of a mystery—it is a sterile triploid, possibly derived from *C. cartwrightianus*, a species from the Greek islands.

As with many other bulbs (or more correctly geophytes, as crocuses have corms), it was the Dutch who pioneered their cultivation; the first record of the plants there speaks of them as an introduction from Istanbul in the 1560s. Various wild central European species were introduced into gardens over the next few centuries, with the arrival of the large and colourful *Crocus vernus* making a particular impact in the 18th century. A number of species were then introduced from the Ottoman lands, including the Balkans (e.g., *C. flavus* and *C. angustifolius*), these usually being distributed from Vienna. Later, around the middle of the 19th century, two species introduced from this region, *C. chrysanthus* and *C. tommasinianus*, ended up dominating crocus breeding and production.

> Our tulip-growers have beaten the Dutch in the quality of their best novelties, and we see no reason why we should not beat them in crocuses.
>
> (Glenny 1855)

Attempts at improving the crocus and making it a florist flower had some impact in the 19th century. A large crocus-growing industry sprang up in Lincolnshire, with many expatriate Dutch working alongside locals, particularly in the area known as Little Holland around Spalding; according to a contemporary observer, "It is not too much to say that at least nine-tenths of the Dutch bulbs which are advertised annually as 'Just imported from Holland,' are from Holland in Lincolnshire, and are guiltless of any connection with Holland on the mainland of the continent of Europe" (G. F. Barrell, *Gardeners' Chronicle*, 1 January 1887). E. A. Bowles made many crocus selections and did much to popularise them during the early 20th century.

Cyclamen Primulaceae

Twenty-three species of *Cyclamen* are found across the eastern Mediterranean region, as far as Iran in the east, Libya to the south, and central Europe to the north, with an outlier in Somalia; the centre of diversity is Greece and Turkey. The name comes from the Greek for "circle," possibly alluding to the round shape of the tubers. Cyclamen are geophytes, or, to be precise, corms, whose period of active growth exploits the cooler, moister growing season of the Mediterranean winter. The plants are generally found in semi-shaded woodland habitats. The seeds are coated in a sweet substance, which attracts ants who carry them off to their nests—the seedlings can then germinate in the weed-free and well-cultivated environment of the ant-heap. The tubers can grow to enormous size; lifespan is unknown but possibly decades.

Cyclamen were occasionally used in herbalism, particularly for aiding in childbirth; it was believed in some places that pregnant women could suffer a miscarriage simply by stepping over the plant.

Cyclamen hederifolium and *C. repandum* began to appear in German gardens in the 16th century, in English somewhat later. *Cyclamen persicum* was established in cultivation by the mid-18th century, but serious selection work did not happen for another hundred years, when growers began to expand the colour range, the first red, 'Wilhelm I', appearing in Germany in 1868. An English grower, Edmonds of Hayes, produced 'Giganteum', a tetraploid, in 1870, and from this developed a range of different colours. In the hands of German breeders work continued apace, with good reds a particularly important goal. By the 1930s there was an interest in smaller varieties, but attempts at crossing *C. persicum* with other species proved singularly unsuccessful, as it has to this day.

Mrs. C. Saunders and Lewis Palmer, two English gardeners, photographed by Valerie Finnis (1924–2006), whose pictures record many of the great horticulturalists of the 20th century. They are in an unusual vintage greenhouse with *Cyclamen persicum* wild form. Courtesy of RHS Lindley Library.

Smaller, more delicate plants were finally successfully produced by a breeding programme run by Wye College in southeast England in the 1960s, by backcrossing modern hybrids with the wild plant. During the 1970s, it was mostly German and Swiss breeders who continued this work; most cyclamen sold now are minis or the slightly larger midis—the public like them, and they are cheaper to transport. This period also saw the development of F_1 strains, important for commercial growers because all the plants would then flower at once and could go to the flower auction together.

Akira Fuse's 1975 pop hit "The Scent of Cyclamen" made the plant immensely popular in Japan, and many breeders are now working there. The use of cyclamen for bedding has become widespread in those areas of Japan with mild winters, as it has become so too across northern Europe since around 2000, largely a reflection of climate change but also the efficient mass production of what is now a highly predictable crop plant.

The late 20th century saw a surge of interest in wild cyclamen. The plants had been much threatened by commercial collecting, particularly in Turkey, until the international CITES treaty entered in force in 1975. Since then, the ease of propagation by seed has resulted in several species, particularly *Cyclamen hederifolium* and *C. coum*, becoming common in the nursery and garden centre trade. Others remain connoisseur plants.

ABOVE LEFT A hand-coloured engraving of a cyclamen, possibly *Cyclamen purpurascens*, from Georg Wolfgang Knorr's *Thesaurus rei herbariae hortensisque*, volume 1, published in Nuremberg, Germany, in 1770, clearly indicating that some at least at this time considered the plants ornamental enough to grow in pots. Courtesy of RHS Lindley Library.

ABOVE *Cyclamen persicum* varieties in a plate from the first volume of *The Illustrated Bouquet, Consisting of Figures with Descriptions of New Flowers* by "the eminent florists" E. G. Henderson & Son, dated 1857. Courtesy of RHS Lindley Library.

LEFT Entitled "The Persian Cyclamen" (i.e., *Cyclamen persicum*), this is an engraved colour plate from Robert Thornton's *Temple of Flora, or Garden of the Botanist, Poet, Painter, and Philosopher*, published in London in 1812. The engraving is by Philip Reinagle. Courtesy of RHS Lindley Library.

Dahlia Asteraceae

A genus of some 40-odd perennials from Central America, most probably named for Anders Dahl, a student of Linnaeus, by Antonio José Cavanilles, director of the Royal Botanic Garden of Madrid, who was one of the first to send seeds to Europe, in 1788. French botanist Thierry de Menonville also sent seeds at this time, and the great German explorer Alexander Humboldt made several collections in 1802–03. A very flexible genetic make-up has resulted in frequent wild hybridisation and allowed for rapid, and indeed spectacular, development of a range of hybrids in cultivation—currently some 57,000 varieties are recognised. The range of colour is very wide (but no blues), as is flower form, with varieties being classified into groups defined by flower shape—or more correctly, since these are Asteraceae, flower heads (often referred to as "blooms" by dahlia growers), composed of multiple florets, each with one petal. Of these groups, the most important are as follows:

- Singles. Only the outer ray florets have petals, with the inner disc florets lacking a petal; generally grown as border plants or as a source of cut flowers.
- Anemone-flowered. The inner florets form a central boss of small tubular petals.
- Collarette. With an inner ring of smaller florets; most are used as border plants.
- Waterlily (Peony). Fully double with a flattish profile.
- Decorative. The most numerous group, with fully double flower heads, the largest of which may be 35cm across—such monsters are very much grown for competitive showing.
- Pompon. Fully double, with in-rolled florets, the whole forming a ball; popular for borders and showing.
- Cactus. Fully double, with pointed and largely in-rolled florets giving a "spiky" appearance.
- Semi-cactus. Similar, but with broader and less strongly in-rolled florets.
- Miscellaneous. Various shapes (including the very popular 'Bishop of Llandaff') that do not fit easily into any other group.

Dahlias are closely linked to oak and pine forest communities of Mexico and Central America between 1,000 and 3,500m in elevation; these regions generally have a rainy warm season and cool dry one. In nature, plants are found in more open areas; now they are common roadside plants. Dahlias are herbaceous perennials, non-clonal but

Collarette (left) and Mignon (right) dahlias from the 1959 Stassen catalogue. Mignons are a group of dwarf dahlias, recommended for bedding displays.

long-lived, which die down in the cool season to a tuber. That this tuber can be dug up and stored over winter in frost-free conditions has greatly facilitated cultivation in cold-winter climates. The plant was grown by the Aztecs and was used medicinally; there is little evidence that they made any distinctive ornamental selections. Always a minor part of the diet, the tubers are still eaten in some regional Mexican cuisines.

The original species behind the garden varieties are *Dahlia coccinea* and *D. pinnata*, both naturally very variable. There was a wave of hybridisation almost as soon as the plants arrived in Europe, with German growers in the lead for much of the 19th century. Plants were initially grown in botanic gardens (Berlin was one of the first), then used in public parks and the gardens of the wealthy. Those further down the social scale who were able to obtain plants found it relatively easy to keep them, making this a more democratic garden plant than many exotic 19th-century introductions. A German source listed 103 varieties in 1817.

The story of dahlias in cultivation is very much one of ups and downs, with fanatic crazes followed by fallow periods. Garten-inspektor Hartwig of Karlsruhe produced a full double in 1808—the first pompon (named for the ball on top of a French sailor's hat). Christian Deegen established a nursery in Bad Köstritz in Thuringia in the 1810s; other nurseries specialising in the plant also set up here, and the town remains a centre for dahlia growing, with its own museum and exhibition ground, the Dahlien Zentrum.

> My advice to the women's clubs of America is to raise more hell and fewer dahlias.
>
> (William Allen White 1868–1944)

In Britain the 1840s and 1850s saw a boom, but by the 1870s interest had dropped off, although the National Dahlia Society was formed at the beginning of this decade. On the European mainland, interest was kept up by the mysterious arrival in 1872 in the Netherlands of the first cactus-flowered dahlia from Mexico, although nothing like it had been recorded in its country of origin. North Americans were relatively late getting into the dahlia cult, with the American Dahlia Society not being founded until 1895.

DAHLIA COCCINEA VAR.

Dynastic connections (i.e., the Habsburg family) between the Spanish and Austro-Hungarian empires facilitated export of seed to Vienna, and thence to the rest of the Habsburg domains. The Czech lands were in any case a hotbed of agricultural and horticultural improvement, and dahlias took off here particularly strongly. One of the most extraordinary stories concerns Božena Němcová (1820–1862), who was crowned "Dahlia Queen" a few days after her wedding in 1837 at the Dahlia Ball, organised by the recently formed Czech Dahlia Society. Němcová went on to become one of the leading lights in the revival of the Czech language and of Czech nationalism, but given the repressive nature of the Habsburg regime, campaigning had to have a cover. Activities to do with growing dahlias became a front for nationalist agitation, a situation commemorated by Smetana in his "Dahlia Polka."

The early 1900s saw continued breeding, and by the latter part of the century, corporate breeders were focussing on compact and dwarf varieties for bedding out and container use. In Britain dahlias had largely become a working-class flower by the 1960s; in the 1990s, however, garden writer Christopher Lloyd (1921–2006) began to champion them, causing a massive reversal in their fortunes. The role the scarlet, dark-leaved 'Bishop of Llandaff' played in this resurgence cannot be overstated; an easy plant with an informal character, it works well when grown en masse but also joins seamlessly with other plants in combination.

OPPOSITE Three varieties of *Dahlia coccinea*, clockwise from top: 'Mrs. Burbidge', 'Bronze', 'Negress'. From *L'illustration horticole* of 1884. *Dahlia coccinea* is one of the wild parents of the cultivated dahlia; it is quite possible that these are hybrids rather than selections of the species.

Daphne Thymelaeaceae

The genus is named for Daphne, the nymph who turned herself into a plant in order to escape being raped by Apollo (never mind that the mythical plant in question is thought to have been *Laurus nobilis*). There are 95 species, all shrubs, mostly evergreen, found across Eurasia, with a few in North Africa. Daphnes are frustratingly temperamental garden plants; they can also be difficult from cuttings, adding to their cost, mystique, and the grief when one is lost. Their unpredictability is possibly linked to their being pioneer plants (so therefore having a distinctly defined lifespan) and an intolerance of root disturbance linked to a general (but not universal) inability to regenerate vegetatively. Their pioneer character is dramatically illustrated by the vast numbers of seedlings of *Daphne bholua* which sprout alongside trekking paths in the Himalaya, although in fact this seems to be one of the few species that can produce clonal suckering growth from the base. "Typical" daphne habitat is open woodland, woodland edge, or scrub, overlaying limestone. Some species (e.g., *D. cneorum*, *D. sophia*) are noted as relict populations, possibly near to a natural extinction.

Flowering tends to be early, with some species moth-pollinated. Though highly poisonous, daphne berries were an ingredient in an 18th-century herbal cure for venereal disease, the Lisbon Diet Drink, but the most useful part of daphne is its bark, which is

[*Daphne cneorum*] shows neither fear nor favour. You may court it in vain, as you may court a cat, with cossetings and comforts uncounted—but with no result to show but the sickliness of a dying plant; yet in some neighbour's garden, where it was ignorantly shovelled into hard common earth to act as an edging like Arabis, it will have run far and wide [with] ample heads of waxy, brilliant, rosy trumpets that fill the air of June with fragrance.

(Farrer 1919)

Daphne cneorum and a *Viola* sp. in the Italian Alps.

used in papermaking in the Himalaya and China. In Nepal *lokta* paper is widely made and highly regarded for its rich fibrous texture; increasingly it is being managed as a sustainable resource and promoted as an important source of livelihood in remote rural areas. In traditional Chinese medicine, the bark of *Daphne tangutica* specifically is used for its analgesic, coronary-dilating, and anti-inflammatory effects.

Daphne odora was grown in Song dynasty China. *Daphne cneorum* was cultivated from the 18th century onwards, in England at least, and *D. mezereum* was sometimes grown as a cottage garden plant, largely for the scent of its very early flowers. Other species were largely introduced into cultivation in the late 19th century. Daphnes have always been seen as rather choice and special plants, with only a few species and cultivars making it into more general cultivation. There has been little hybridisation.

Delphinium Ranunculaceae

Delphinium is a complex herbaceous genus with many taxa showing a superficial similarity to each other; the overwhelming majority of the more than 300 species are found in eastern Asia, with others in Europe, North America, and high-altitude areas in Africa. California has a large number of species. The genus includes annuals, biennials, and rather short-lived perennials. Most are cool-growing or spring-growing plants, flourishing in mountain or higher-latitude climates, with summer dormancy often the first reaction to heat or drought; those in warm-summer climates tend to die back early to tuberous roots. Most are found in light woodland, tall-herb flora, or grassland. Plants in cultivation have persisted well over 10 years, although they do not appear to be true clonal plants.

Delphinium elatum in the wild, in steppe grassland, Kyrgyzstan.

"New Delphiniums" by Ludwig Bartning (1869–1956), a German painter of flowers and still lifes, in Foerster's *Gartenstauden Bilderbuch* (1938), from left to right: 'Fon', 'Rose Quartz', 'Tropical Night', 'Lighthouse', 'You Blue Wonder', 'Glacierwater', 'Eyecatcher', and 'Goodnight'. Courtesy of Bettina Jacobi.

Flower colour is various, including strong reds, many pale cream tones, and that rarest of flower colours, intense true blue—in great columnar masses; breeders in many countries have worked hard to improve the delphinium, these natural gifts notwithstanding, as much for the commercial cut flower trade as for garden use. The name comes from the Greek, for the seeming resemblance of the bud to a dolphin. Greek legend tells of the springing up of delphiniums from the spilled blood of the hero Ajax; various cultures have used parts of the plant as a dyestuff—blue, but also yellow. The plants are generally toxic, and there are few herbal or other uses.

Clusius described *Delphinium elatum* in 1601. This is the core species, with the modern flower the product of progressively adding others, such as the Siberian *D. cheilanthum* and the Chinese *D. tatsienense*, both of which Lemoine brought into the gene pool. James Kelway was the first breeder in the United Kingdom, acquiring stock from

KELWAY'S NEW DELPHINIUMS.

SHOWING THE ACTUAL SIZE OF THE INDIVIDUAL FLOWER PIPS.

LOVELY (Kelway) 21/- each.
HUISH BEAUTY (Kelway) 10/- each.
BEAUTY (Kelway) 3/- each.

SILVER BUCKLE (Kelway) 10/- each.
REMARKABLE (Kelway) 10/6 each.
F. CARR (Kelway) 10/- each.
PERSIMMON (Kelway) 3/6 each.

SIR J. G. WARD (Kelway) 10/- each.
EVENING (Kelway) 5/- each.
KING OF DELPHINIUMS (Kelway)
3/6 each.

Individual flowers of delphinium hybrids in *Kelway's Manual* of 1924. Courtesy of Kelways Plants Ltd.

France from 1859 onwards. In Germany, Foerster was famous for his delphiniums; the flowers had a particular appeal in the early 20th century in that country because of the mystical appeal of the colour blue; artists, printers, and designers were fascinated by it and its pure representation. From 1904 the English firm of Blackmore and Langdon produced some of the finest new varieties, and continues to do so. Modern varieties are semi-double, where the flower has 13 sepals with a reduced spur, and at least four petals in white, black, or a self colour. These semi-double flowers last longer than the singles found in the species or older hybrids. In good modern hybrids, the youngest flowers at the base are still alive as the youngest ones at the apex are opening.

Frank Reinelt (1900–1979), a Czech immigrant to California and once head gardener to Queen Marie of Romania, developed the Pacific strain in the United States during the 1930s, working by alternating generations of hand pollination with open pollination. He sought to produce plants for a wide range of climatic conditions; in many U.S. climate zones delphiniums are particularly short-lived, so seed strains are more useful than the named cultivars favoured in Europe. He later worked on what has become something of a holy grail for breeders—red and pink flowers, initially using the scarlet *Delphinium cardinale* to produce the Astolat and Elaine series. The Dutch breeder B. Ruys, working in the early 20th century, was also something of a pioneer with these, but using *D. nudicaule*.

Deutzia Hydrangeaceae

Named for Johann van der Deutz, an 18th-century Dutch lawyer who contributed to one of Thunberg's expeditions to Japan. Around 60 species of these deciduous shrubs are known, 50 of them Chinese. The rest are found across Eurasia, with some also in Central America. Many of the species have a fatal weakness, in that they get in to growth early and are consequently damaged by frosts. Deutzias are found in light woodland and scrubby habitats, often at altitude. They are true shrubs, constantly regenerating from the base. The wood is hard and has been used for cabinet-making in eastern Asia. The leaves of *Deutzia scabra* are coarse enough to be used for polishing.

A few species were introduced to cultivation in the 19th century; Vilmorin received many early collections, from Delavay and others. The early growth of *Deutzia gracilis* was put to good use immediately, as it made it an ideal shrub for forcing; large numbers of plants were grown in Holland, in vacant spaces in the bulb fields, and then exported to Britain for this purpose. Lemoine made the crosses that effectively brought the plant to popular attention; many of his hybrids, such as 'Mont Rose' and 'Magicien' are still common. Good forms continue to be introduced to this day.

> These shrubs deserve a better fate than that of the common shrubbery, mixed up with all sorts of things of different natures and sizes. [They] should not be reduced to mopheadedness by cutting back [and it is] better not to prune at all than to hack them into ugly shapes.
>
> (Robinson 1921)

Dianthus Caryophyllaceae

The name for the genus, which includes the various flowers known as pinks, carnations, and sweet William, is derived from the Greek, *dios* ("divine") and *anthos* ("flower"). The 320-odd species are found across the temperate regions of the northern hemisphere, with a centre of diversity in the region around the northern shores of the Mediterranean, including the European Alps. Plants favour dry rocky places, with a distinct preference for limestone: several species have become naturalised on castle walls in northern Europe, in some cases probably several centuries ago. All are either herbaceous or extremely weakly subshrubby, in that there is persistent above-ground overwintering growth, although it is often very soft. In some of the Turkish species of alpine or semi-arid environments, this growth takes the form of a cushion, which in some species may be very hard. These latter species may be very long-lived, whereas plants from the gene pool in general cultivation are best characterised as short-lived perennials. Most species can be seen as being on a gradient between stress-tolerators and pioneers.

Traditionally, there is a strong mythical connection with eyes. When the Greek goddess Artemis, for example, shot out a young man's eyes with arrows, dianthus plants sprang up from the pair as they lay on the ground, while in Christian tradition they (like *Convallaria majalis*) are said to have arisen from Mary's tears at the Crucifixion. Pinks have been used for flavouring wine, although they have little flavour to impart. And carnations have always had strong associations with marriage (in many European countries, they are a typical wedding boutonniere) and motherhood (as with their use on Mother's Day—or the anniversary of Roe v. Wade—in the United States). Red carnations specifically have been used as a symbol of the socialist movement; Portugal's Carnation Revolution of 25 April 1974, however, was a more general revolt against a fascist regime by progressive army officers, who inserted red carnations into their gun barrels to signify their peaceful intent.

Alpine growers cultivate a great many *Dianthus* species, while a few others have played a major role in garden history. Three species in particular have a history in European cultivation which appears to go back to the Medieval period: *Dianthus barbatus* (sweet William), *D. caryophyllus* (carnation), and *D. plumarius* (common pink). The contemporary gene pool is the result of centuries of crossing—spontaneous in the early period. A fourth species has also had a major impact—*D. chinensis* (the so-called Indian pink), introduced from Japan via Russia in

A carnation-type pink, (possibly) by Jacob Marrel (1614–1681), a German painter of flowers and still lifes who worked in the Dutch provinces. Courtesy of the Rijksmuseum, Amsterdam.

1860. And *D. knappii*, from eastern Europe, has contributed genes for yellow flowers.

Gillyflowers, as *Dianthus* species were commonly and historically known, were renowned for scent, an important aspect in the Medieval period, when foul smells must have been abundant. Additionally, it was believed that bad odours caused disease, including the Black Death, so sweet fragrances were seen as having a protective effect. Paintings from 1500 onwards show plants in pots, and literary references increase from this time. The biennial *D. barbatus* was grown extensively in Tudor times but was then displaced by derivatives of *D. caryophyllus* and *D. plumarius*. There was a great deal of diversity—John Rea (d. 1681) records that there were 360 varieties.

Laced pinks (derived from *Dianthus plumarius*, and generally white with a contrasting eye and a clearly defined zone of the eye colour around the petals) were developed in the 18th century; James Major, gardener to the Duchess of Ancaster, is credited with producing the first, c. 1770. These were taken up as a florist flower and grown competitively, particularly by weavers. Some flowers were so large (10cm across) that they needed glued pigs' bladders and other devices to support them. Later revivalists of the plant made a strong association with the Scottish city of Paisley, now questioned by historians, although the Paisley Florists' Society was undoubtedly large and well organised.

OPPOSITE Two "perpetual" carnations decorate the front cover of the 1928 Vilmorin-Andrieux catalogue. The Vilmorin family company dominated French agriculture and horticulture from 1743 until it was sold in 1972; it is now part of the Limagrain group.

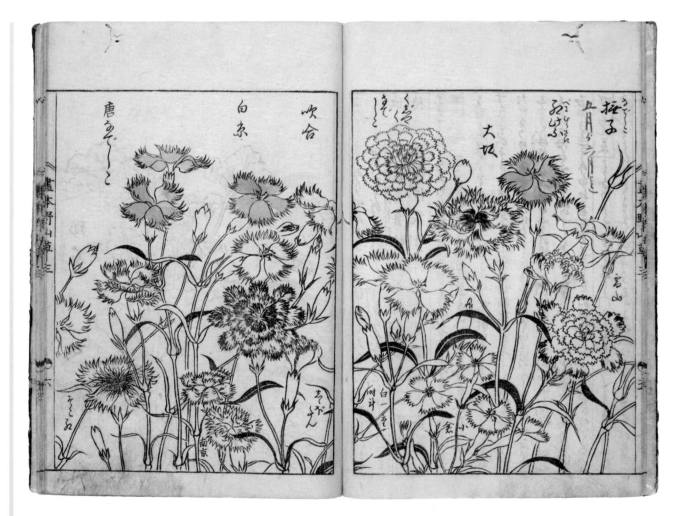

Cultivars of pinks in *An Illustrated Book of Wild Flowers*, published in Japan in 1755, along with a note that alien plants are included as well. Courtesy of Chiba University Library.

1928

Vilmorin-Andrieux & Cie

4 Quai de la Mégisserie,

PARIS (1er)

BERTAULT

ŒILLETS — 1. ŒILLET PERPÉTUEL GÉANT. — 2. ŒILLET PERPÉTUEL GÉANT ENFANT DE NICE.

The carnation has somewhat mysterious origins; almost certainly it was introduced from the Ottoman Empire, and its spread across Europe (as evidenced by artistic representation) was rapid. The malmaison carnation, bred initially in early-19th-century France, has a very large double centre of small petals; later in the century it became the signature of a good head gardener to grow on a malmaison for four or five years to produce a plant with 50 to 100 flowers. By the 1830s, certain new varieties of carnation bloomed almost continually. Sim carnations, scentless but hugely important commercially for cut flower trade, were bred by William Sim, a Scots emigré to the United States in the early 20th century.

Dianthus superbus, whose range nearly spans the breadth of Eurasia, is distinctive for its deeply fringed petals. In cultivation, various forms have been named, and it has entered the general pink gene pool. *Dianthus longicalyx* has a long history of cultivation in China and Japan, where it has been grown since the Heian, its autumn flowering and easy hybridisation and cultivation making it popular but also a plant with a lot more mass appeal than many of the cult plants of the elite. Forms classed as *ise-nadeshiko* have incredibly long fringes on the petals, while *tokonatsu* flower all through the summer. At the end of the Edo period, Western varieties joined the gene pool; the result was a surge in their fashionability in the 1920s.

Along with the rather industrial Sim carnations, the 20th century's greatest breakthrough was made by Englishman Montagu Allwood (1879–1958), who set up a nursery specialising in *Dianthus* in 1910; he spent nine years crossing the perpetual-flowering carnations with old-fashioned border pinks to produce hardy varieties that bloom all summer. Allwood spent the rest of his life breeding pinks, the best being 'Doris', which can continually produce crops of perfect flowers—up to 440 stems per square metre per season under glass.

'Youell's Duke', from the *Florist's Journal* of 1840. As with many cultivars of the period, the name commemorates the name of the breeder and adds an aristocratic flourish. Courtesy of Sophie Hughes.

Catalogues abound in wrongly-named Dianthus, in undescribed or indescribable or merely mongrel Dianthus.

(Farrer 1919)

Digitalis Plantaginaceae

The 22 species of *Digitalis* (Latin for "finger-like") are found from Europe across to central Asia, with some in northwestern Africa and Australasia. All are short-lived perennials or biennials, or in the case of some species of warmer climates, weakly woody subshrubs. The herbaceous species are very much plants of woodland edge habitats, acting as pioneers after felling or wind damage, or growing in odd glades where light levels permit. The seed can remain buried for years, germinating when the soil is disturbed. Some Mediterranean and Middle Eastern species are plants of rocky places and other open situations where good conditions for growth may be relatively short-lived. The genus was traditionally placed in the Scrophulariaceae, but recent phylogenetic research has placed it in the now greatly enlarged Plantaginaceae.

Plants are commonly called foxgloves, after the most well-known species, *Digitalis purpurea*, the source of the drug digitalin, used in treating heart problems. But early herbalists had no clue of this: Parkinson wrote in his *Paradisi in Sole Paradisus Terrestris* (1629), "Foxegloves are not used in Physicke by any iudicious man that I know." The effect of digitalin was not discovered until the latter years of the 18th century, and the plant then became so important that apothecaries in France would hang a painting of the flower on the outside of their shops to advertise their skills.

Digitalis purpurea made the transition from herbalist's to cottager's garden as much through its ability to naturalise as perhaps any conscious decision. It has a strong association with the faerie folk in Britain; sometimes this meant that foxgloves could be used protectively, at others that they should be kept out of the home for fear of bringing in malign influences. From the latter years of the 19th century, selections of the species produced seed strains that were predominantly white or some other colour variant, and over the course of the 20th, other European species became more frequent in gardens. Most commercial breeding still focuses on *D. purpurea*, producing seed strains that once established as a self-seeding population in a garden rapidly revert to plain pink.

Echinacea Asteraceae

Echinacea includes nine species, commonly called coneflowers, their generic name derived from the Greek for "sea urchin," after the shape of the spiky seedheads. All are herbaceous perennials found in eastern and central North America, plants of prairie habitats and open woods. *Echinacea purpurea*, however, is more a plant of moist savannah than true prairie; it is not long-lived in the wild. Other species are longer lived, but none are clearly clonal perennials, as they tend to form tight clumps with limited vegetative spreading. Lifespan in cultivation is affected by many factors, with hybrids often notably short-lived.

Echinacea species were widely used medicinally by Native American communities, often for common cold–type ailments. This led to them becoming popular in folk medicine among American settlers during the 19th century—a use which in the 1990s experienced a major revival; however, there is little evidence that any of the species (or preparations developed from them) have much effectiveness.

> It is extremely frustrating to have to admit that *Echinacea* is not reliable.
>
> (Gerritsen and Oudolf 2000)

Echinaceas appeared in German and English gardens in the last two decades of the 18th century. The plants, most commonly *Echinacea purpurea*, remained in cultivation as unimproved species until the latter part of the 20th century, when a number of selections were made by nurseries, followed by a flurry of hybridisation, the key aspect of which was incorporating the yellow from *E. paradoxa* into a wider gene pool; the genus is in many ways one of the most "modern" of garden perennials. The plant has also become something of a poster child for the native plant movement in the United States.

Epimedium Berberidaceae

The 54 species of *Epimedium* are dominated by the 43 from China, the remainder being found in the Caucasus and around the Mediterranean, a good example of a disjunct distribution as a result of the ice ages breaking up a once much wider distribution. The name is derived from a Greek plant name, but for what is uncertain. The North American *Vancouveria* is very closely related. Epimediums are all long-lived perennials, often evergreen in mild winters, and are generally from forest and scrub in temperate mountain areas. They are forest-floor stress-tolerators that spread clonally, with persistent growth, eventually forming large patches which effectively exclude other species. The Western species have a long history in cultivation and can be used as ground cover; the more recently introduced Chinese species seem less vigorous. A distinctive feature of the flower are the long horn-like nectaries of the petals, clearly designed for specific pollinators, but which may be punctured by bees to access the nectar. The seeds are distributed by ants.

Epimediums have a long history in traditional Chinese medicine, partly as an aphrodisiac (and indeed, it has been discovered that they contain a compound very similar to the active ingredient in Viagra, with a similar effect). As a consequence, the widespread digging of plants has had a severe impact on many wild populations. Other medical uses are being investigated.

Tournefort was the first botanist to define the genus, in 1694, although Gerard was growing one in London a century before. Von Siebold brought several more from Japan in the 1820s; the species from China were unknown to Europeans until Maximovich's travels in the 1870s. The 20th century saw several selections and hybrids developed from the Western species, and these were used, sometimes quite extensively, in gardens and other designed landscapes. Since the 1980s, many more species have been introduced from China; many of these have very decorative flowers, and they also hybridise readily, factors which are making them increasingly popular garden plants.

An illustration from the 1921 edition of William Robinson's *The English Flower Garden* showing how epimediums were sometimes used in the late 19th and early 20th century for forcing to provide early foliage for the home.

Erysimum Brassicaceae

Erysimum (derived from the Greek name of the plant) now includes the familiar wallflower, *E. cheiri* (formerly *Cheiranthus cheiri*), which has been a frequent feature of old and ruined walls for centuries across Europe, possibly for millennia, although originally native to the eastern Mediterranean region. The 180-odd species are found from western Asia (the centre of diversity) and across Europe to North America, but no further south than Costa Rica. The plants vary from annuals to short-lived subshrubs, with a very weakly woody element; they are pioneer plants in habitats where other vegetation is open enough to allow for reseeding, as well as rocky places and scree.

Traditionally a sprig of wallflower was worn in the caps of wandering minstrels to commemorate a (Scottish) princess who fell to her death from a tower on trying to elope—her bereft lover became a well-known wandering minstrel.

Wallflowers were cultivated in Europe during Classical times, while in the Medieval period they are known from the 13th century onwards. Doubles were recorded and maintained as cultivars from the late 16th century, with the best forms potted up and kept under cover in the colder parts of central Europe. During the 18th and 19th centuries an increasing number of distinctly coloured cultivars were maintained; often these were grown as windowbox plants or as sources for cutting. Being easy from both seed and cuttings, they became popular with all classes.

Basic garden wallflowers are now generally grown from seed strains, but some varieties, such as the historic (1880s) *Erysimum cheiri* 'Harpur Crewe', are maintained vegetatively as cultivars. There is a wider gene pool than the *E. cheiri* types, and these too are maintained by vegetative propagation. The best-known example is *E.* 'Bowles's Mauve', thought possibly to be an *E. linifolium* hybrid; its name honors E. A. Bowles, of course, but there is no record of his having grown it.

OPPOSITE Two wallflower varieties from *Beautiful Garden Flowers for Town and Country* (Weathers 1904), illustration by John Allen.

Eschscholzia Papaveraceae

To answer the first question that always come up with what are commonly known as California poppies, this genus of short-lived plants is named for early-19th-century Russian/Estonian naturalist Johann Friedrich von Eschscholtz. All 10 species are found in the western United States and Mexico. Eschscholzias are annuals or short-lived perennials, the well-known *Eschscholzia californica* being a typical winter annual of Mediterranean climates, often living from first autumn to second winter in well-watered climes. Plants flourish, as do all such annuals, in terrain where the existing vegetation canopy is open enough to permit seedlings to survive. The first reports of California named it a "land of gold" because the plants coloured the hills so intensely with their flowers. The settlers brought in turf-forming grasses which smothered much of their habitat, as well as that of many other species. Elsewhere, plants naturalise easily and can become invasive, as they have in Japan.

The plant contains alkaloids, related to but weaker than those found in opium poppies. It was widely used for treating a range of ailments by Native Americans and remains popular with herbalists for its calming effect, for example in treating attention-deficit disorder. This usage has some, cautious, scientific support.

Eschscholzia californica was discovered by Europeans in the late 18th century with seed being sent back soon after. It was popular as an annual across Europe by 1840. Selection of different colour forms and of doubles took place during the latter part of the century, mostly in Britain. Selection of seed strains has continued ever since, and the plant can certainly be regarded as an important part of the annual garden flora, although its breeding has never attracted a great deal of investment.

> Vast fields of Golden Poppies have ever been one of the strong and peculiar features of California scenery. The gladsome beauty of this peerless flower has brought renown to the land of its birth.
>
> (Emory E. Smith, preface to *The Golden Poppy*, 1902)

Euphorbia Euphorbiaceae

The name honors Euphorbus, a physician in North Africa in Classical antiquity who was noted for his contributions to natural history, including notes on these plants. With around 2,000 members, *Euphorbia* is among the largest genera of flowering plants, with species varying from small trees and shrubs through to perennials and annuals. Distribution is worldwide. There is a strong tendency for euphorbias to flourish in dry places, with around half having a distinctly xerophytic character. Some African species are an interesting example of parallel evolution, as at first sight they look just like cacti. The hardy species grown ornamentally are all members of the subgenus *Esula*, the least evolved of four subgenera, with a centre of diversity in the Mediterranean region. These tend to come from drier habitats, from rocky slopes to seasonally dry woodland. Nearly all are clonal perennials with competitive and stress-tolerant characteristics, with some exhibiting guerrilla spread, the rate of which can vary greatly between species. The well-known Mediterranean-origin *Euphorbia characias* is interesting on two points: one that it is a rare example of a cultivated herbaceous plant which is effectively evergreen, as new stems replace the old ones just as they are dying in an annual cyle; the other is that it is non-clonal and short-lived, with a distinct pioneer character. *Euphorbia* is the only genus to have all three photosynthetic pathways: CAM, C_3, and C_4.

Euphorbias have a unique flower structure—stripped down to a bare minimum, but with some new structures added—an evolutionary one-off so surprising that looking into a euphorbia flower can be a disorienting experience. Male and female flowers are separate, with the male reduced to little more than one stamen, the female to a style with three fused carpels, each of which is capable of producing a seed. These parts are each enclosed in a cup-like cyathium, which is a completely novel structure unique to the genus. There are nectar glands at the base of this and then some cyathium "leaves," which in turn are surrounded by bracts (modified leaves). The latter two sets of structures give the "flower" its colour—nearly always yellowy in hardy species but often red in those from warmer climates, such as *Euphorbia pulcherrima* (poinsettia).

A milky, acrid, latex-like sap is common to the whole genus and a potent deterrent to predators—smaller ones get their mouth-parts glued together and larger ones get burnt. Humans vary in the sensitivity of their skin, but great care must always be taken to avoid getting it into the eyes, as extreme pain and temporary blindness may result. The chemically complex and toxic sap of *Euphorbia lathyrus* (molewort) can be used to burn off warts and moles (and it follows that the common name has led to the myth that growing it deters garden moles). In 20th-century Mexico, *E. antisyphilitica* (candelilla) was used to produce a wax that served various purposes, e.g., waterproofing and polishing, but has largely been replaced by petroleum-based (and probably safer) alternatives; its scientific name indicates a former use in herbal medicine—probably highly risky!

> In spring, [*Euphorbia pilosa*] and *E. amygdaloides* are attractive by their yellow flowers when little else is in bloom, but they are scarcely worth growing in a general way.
>
> (Robinson 1921)

Of all our contemporary perennial garden flora, this is the last major addition; the Robinson quotation (see sidebar) well indicates that the appeal of the genus is historically quite recent. *Euphorbia lathyrus* has been known since the Medieval period. Records of *E. myrsinites* and *E. epithymoides* (as *E. polychroma*) appear somewhat later. Silva-Tarouca mentions several, as does Robinson. The plants stayed overwhelmingly in botanical and enthusiasts' gardens until the 1950s, when *E. characias* and *E. epithymoides* led the field in making their way into gardens. It is probably significant that the widely read garden columnist Margery Fish wrote about euphorbias in the 1960s; the plants she discusses are very much the ones available today, plus a few new faces such as *E. schillingii* (from Nepal, 1975) and a growing range of *E. characias* cultivars.

Ferns

Ferns are very different from flowering plants. They arose around 360 million years ago in the late Devonian period; this does not necessarily make them "living fossils," however, as most of today's fern families evolved and dispersed during the early Cretaceous (around 145 million years ago), concurrently with many flowering plant families. This early evolution led to very wide distribution across the continents. Depending on definition, there are around 9,000 species of true fern, currently divided into 40-plus families. Ferns in cultivation tend to be represented very unevenly, almost arbitrarily, with many distinctive and interesting plants rarely seen, even in botanical gardens. The following are the most horticulturally relevant:

- *Adiantum* (maidenhair fern) is named from the Greek for "not wetting," referring to the fronds' hydrophobic qualities. It is a genus of about 200 species, found typically in moist shaded environments, with a strong propensity to flourish on wet cliffs, water seepages, and around waterfalls. Generally clump-forming, sometimes running. A global distribution with centres of diversity in the Andes and China.
- *Dryopteris* (Greek for "wood fern") has about 250 species across the temperate northern hemisphere, with the greatest diversity in eastern Asia. They are mostly woodland plants that form a long-lived, slowly expanding crown.
- *Nephrolepis exaltata* (Boston fern) is one of the 30 species in its genus (named for its kidney-shaped scales), with a wide distribution in tropical regions throughout the world. This taxon is possibly the most widely cultivated fern, due to its tolerance of drought and light, a general feature of the genus.
- *Polystichum* (Greek for "many rows") is a genus of about 260 species of ferns with a global, largely temperate, distribution, again with the highest diversity in eastern Asia and the Andes. They are plants of shade but with some in alpine regions. Growth habit is similar to *Dryopteris*.
- *Pteris* (Greek for "fern") has one species, *P. cretica*, which is among the most commonly grown of all ferns. The 280 species are found in tropical and other frost-free climates across the globe.

> It is through ferns that one can describe the essence of shade.
>
> (Foerster 1957)

Plants were among the favourite subjects for early experimental photography (they stay still better than animals). This fern image produced by the cyanotype process by an unknown English photographer dates from the very early years, 1840–1845. Courtesy of the Rijksmuseum, Amsterdam.

One of the defining characteristics of ferns is a very clear "alternation of generations"—the fern plants we are familiar with produce spores (asexually), which then germinate on a damp surface to produce a prothallus, a tiny structure in which sexual reproduction takes place, resulting in the production of a recognisable fern plant. The need of the prothallus for humidity is one of the factors that limits the spread of ferns through a given habitat.

The stereotype of ferns growing in damp and shady places has (as is often the case with stereotypes) more than a grain of truth, as this is indeed the typical habitat; however, ferns are fundamentally stress-tolerators, and in many cases have the ability to safely go into dormancy during drought and to rehydrate very quickly and resume normal life—often far more effectively and with less tissue damage than is the case with flowering plants. Consequently many ferns are epiphytes, such as the bizarre *Platycerium* (staghorn fern) or *Asplenium nidus* (birdsnest fern), or lithophytes; it is perhaps surprising that so few others have entered cultivation. The subfamily Cheilanthoideae consists of six genera of desert ferns, generally found growing in rock crevices in dry regions and superbly adapted for seasonal drought; others occupy similar habitats in the temperate zone. Some are grown by rock garden enthusiasts.

Terrestrial temperate zone ferns are always long-lived, usually deciduous or semi-evergreen. Clonality varies; in some cases (notably *Dryopteris*) a stump will build up over many decades, bearing an uncanny resemblance to the trunk of a tree fern, whereas in others there is a strong guerrilla running habit, most dramatically seen in temperate zone gardens in *Onoclea sensibilis*.

There are climbers (e.g., *Lygodium* spp.) and the so-called tree ferns. The latter are frequently seen in subtropical and montane tropical envi100ments and can form a substantial part of the plant community biomass. They do not produce woody growth, as the trunk is composed of dead stem matter ramified by roots, which absorb moisture through the stem rather than from the ground; so the plants are often sold as "logs," lacking any kind of roots. *Cyathea* and *Dicksonia*, from temperate zone southeastern Australia and New Zealand, are now common horticulturally.

Ferns in temperate zones seem largely unbothered by predators; one reason must be that in many cases they are full of toxins. They have attracted relatively little folklore, and uses are few. One, of historical and conservation interest, was the digging up of *Osmunda* species to use their very tough roots as potting compost for epiphytic orchids—a practice that now seems utterly bizarre and another reminder of the devastating impact that the 19th century's obsession with orchids had on the natural world. Many ferns can be eaten as a spring vegetable and are widely collected or even grown for this purpose in the Far East; however, this practice has been linked to the high rates of stomach cancer in this part of the world, as many ferns contain carcinogens.

Ferns have always tended to be of minority interest. The exception was 19th-century Britain, when the Victorians indulged in an episode so obsessive that in 1855 writer Charles Kingsley coined a term for it: pteridomania. Ferns fitted into the Victorian aesthetic, an odd mix of the ornamented and subdued; fern motifs were used on a huge scale during the latter half of the century.

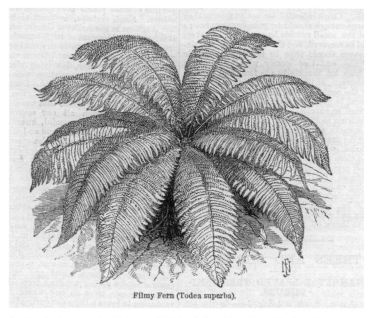

Filmy Fern (Todea superba).

Leptopteris superba (Prince of Wales's fern) is a New Zealand species with fronds up to a metre long. It's an example of a "filmy fern"; these have extremely thin leaves and flourish only in very humid environments. They are not grown at all today but were not uncommon in Victorian Britain, an example of the period's fern fever.

No doubt one factor in their popularity was that ferns flourished in the shaded spaces around the vast numbers of houses being built in the new industrial cities. Enthusiasts collected both species and variations from the norm, the result of which was the wholesale extinction of some taxa from areas of the British Isles, *Osmunda regalis* (royal fern) being one of the most notable casualties. Ferns suffered in the 20th-century reaction against Victoriana and in Britain were almost unobtainable from nurseries again until the 1980s. In Germany, Foerster played a role in helping popularise them in the early part of the 20th century, while in the United States, an interest in growing ferns has been part of the great increase in horticulture generally since the 1970s.

It is possible to create hybrids of ferns, despite their very different method of sexual reproduction. Very few are of importance, although one deserves to be mentioned, *Platycerium* 'Lemoinei', named for and created by that most adventurous hybridiser of all.

Fritillaria Liliaceae

The approximately 100 species of *Fritillaria* are found in three distinct regions. One is centred on the eastern Mediterranean and the Middle East, with one of its outliers, *F. meleagris*, bucking the trend in the genus and having a very wide distribution—right across Europe. Western Asia has a great many species, while the western ranges of North America have 19 species. One, *F. camschatcensis*, is found from the Canadian Pacific coast up round the Bering Strait to Japan; its range hints at it being relatively tractable in cultivation. Many have limited distributions, and are often threatened in the wild by overgrazing or land development. The genus has been relatively botanically stable (although the dramatically dark *F. persica* has been removed from the genus and put back in no less than four times); it is closely related to *Lilium*, sharing with it the rather vulnerable bulb, which lacks the robust skin of many other geophytes. Fritillaries are spring ephemerals, plants of open terrain or sparse woodland, growing in soil which is often moist, even wet, but always well drained, and very often dry for much of the year after a short growing season. They have a reputation for being difficult to grow and have attracted a following of devoted growers, who often refer to them as frits.

> [A]n enormous number of Fritillaries have more or less stinking bells of dingy chocolate and greenish tones.
>
> (Farrer 1919)

Fritillaria imperialis in Iran, Bakhtiari Province.

Fritillaria Meleagris.

KIVITS EYEREN.

Fritillaria imperialis (crown imperial) thoroughly deserves its name. Its extraordinary shape and air of regal magnificence have captivated artists since its introduction to Europe (initially into Italy) from the Ottoman Empire in the mid-1500s; and much folklore is attached to it. Drops of nectar sit at the base of the petals, and various cultures have developed legends to explain this detail over the centuries, including a Persian story about a doomed queen and a Christian one about the tears of a plant which failed to bow to Jesus in the Garden of Gethsemane. Its Western spread was rapid, as so many wanted to get hold of this dramatic new plant; Clusius was involved in developing propagation techniques. Some 13 varieties were known by 1770, with colours that attracted a variety of descriptions from commentators ("dull orange" and "boiled lobster" among them). The plant is the only one of the frits occasionally seen in public formal planting schemes.

Fritillaria meleagris has always fascinated, for the chequered pattern on the petals, but the dull purple-pink has also been seen as a rather sinister colour as the old English names illustrate: sullen lady, madam ugly, widow veil, and snake-flower. It was notably popular in Baroque-era Germany and Austria. Of the great many fritillaries with mysteriously coloured flowers, *F. michailovskyi* is notable for having been bulked up by the Dutch bulb industry through micropropagation and widely sold. The rest are mostly connoisseur plants, although some will not only flourish but even naturalise in gardens if conditions are to their liking. There has been plenty of variety selection in a few species, notably *F. imperialis*, but little hybridisation.

OPPOSITE *Fritillaria meleagris* by Van Eeghen (Blankaart 1698). The super-imposing of the plant on a landscape is typical of the period. Courtesy of the Rijksmuseum, Amsterdam.

Fuchsia Onagraceae

There are just over 100 species in the genus, whose name honors 16th-century German botanist Leonhart Fuchs. All are woody, the vast majority native to South America, the rest to Mexico and the Caribbean, New Zealand, and some Pacific Islands. Horticulture has produced around 12,000 cultivars from less than a handful of original species. All would appear to be potentially long-lived shrubs or occasionally small trees, typically found in montane or cooler and moist-climate environments, in woodland edge and scrub habitats, sometimes above the treeline, or in woodland understorey. They are mostly hummingbird-pollinated in nature. They have naturalised in southwestern Ireland, where they are used for hedging. The berries are edible, if not hugely tasty, and can be made into jam.

The introduction of the relatively hardy *Fuchsia magellanica* (from the temperate southern tip of South America) is shrouded in mystery. There is a good story about nurseryman James Lee seeing a plant in the window of a humble house in London sometime around 1790, and only acquiring it from its owner for a considerable sum (and the offer of replacement plants after propagation). Doubt has been cast on this; it's even hinted that Lee concocted the tale to cover the reality: he stole the plant from Kew.

FUCHSIA FULGENS

Nat. size

PL. 107

What is very well known is that *Fuchsia magellanica* and the more tropical, and therefore tender, *F. fulgens* (from Mexico, 1837) kick-started one of the most prolific of the Victorian era's breeding frenzies. Fuchsias are easy to grow and to propagate, and they hybridise readily. With their exotic good looks and strong colours, the public could not get enough of them—in Britain and indeed in all other industrialising European countries. A French book in 1848 listed 520 varieties; 40 years later around 1,500 were listed. As many as 10,000 plants a day were sold at Covent Garden Market around this time. Vast numbers of them were used to clothe pillars at exhibition centres like the Crystal Palace, among other mass effects. Fuchsias were also popular as a cut flower—the pendant flowers were ideal for the elaborate tiered vases the Victorians were fond of.

Inevitably a decline set in, and by the end of the century many varieties were extinct, the First World War having finished off many nurseries. California, with its mild climate, became a safe haven for some lucky cultivars. The founding of the American Fuchsia Society in 1929 and the British Fuchsia Society in 1937 stemmed the decline; a major task for both bodies was rescuing as many old varieties from extinction as possible. Breeding and commercial production have both continued, but the plant is something of a shadow of its former self. Enthusiast growing in Britain is largely a northern English and working-class hobby.

OPPOSITE *Fuchsia fulgens*, shown in an illustration from Edward Step's *Favourite Flowers of Garden and Greenhouse*, published in 1897 by Frederick Warne & Co. in London. This species has been one of the most important in the development of the cultivated fuchsia gene pool.

Galanthus Amaryllidaceae

The 20 species of *Galanthus* (Greek for "milk" and "flower") are the number one har-
bingers of spring in many temperate gardening cultures. Wild species of snowdrops
are found from the Caucasus, across southern and central Europe, and down into the
Levant. Species tend to be very similar bulbous perennials, largely from woodland habitats. They spread clonally; and indeed given their early season of flower, this may be
regarded as important for their reproduction—many are distributed by animal digging
and consequent movement. The plants are highly persistent once planted and have
naturalised well north of their natural European range. In his 1633 revision of Gerard's
Herball, Thomas Johnson wrote, "These plants [he calls them "bulbed violets"] doe grow
wild in Italie and the places adjacent, notwithstanding our London gardens have taken
possession of them all, many years past. . . . Some call them also Snowdrops" (quoted
in Slade 2014). It is possible that the common name
came from a type of drop earring fashionable in the
16th and 17th centuries.

Compare British populations of snowdrops to
those from mainland Europe, and it is evident that
the plants come from many different locations. It is
thought that *Galanthus nivalis* was introduced to Britain by the Romans, or possibly by the Normans. What
is certain is that snowdrops have been in cultivation
for around 2,000 years, and that large populations are
often connected with monastic sites. The spread of
snowdrops in northern Europe was almost certainly
linked to their popularity as altar decorations at a season when little else was in flower. The pure white, a
colour associated with the Virgin Mary, seems all the
more appropriate for a flower that often blooms on
Candlemas Day (2 February)—the Feast of the Purification of the Virgin.

The flowers were popular as a decorative emblem during the British Victorian period,
when a strong association was made with chastity. The flower was used in church-led
campaigns to improve the behaviour and police the sexual mores of working-class girls;

> We, the Members of the Snowdrop
> Band, sign our names to show that
> we have agreed that wherever we
> are, and in whatever company, we
> will, with God's help, earnestly try,
> both by our example and influence,
> to discourage all wrong conversation,
> light and immodest conduct and the
> reading of bad and foolish books.
>
> *(Proceedings of the Public Morals Conference Held in London on the 14th and
> 15th July 1910)*

there was *The Snowdrop*, a magazine that functioned as a morality comic, and even a Snowdrop Band, a club to promote chastity among factory girls in the northern mills (see sidebar for their membership card). The bulbs contain galantamine, a compound used to treat polio in Bulgaria during the 1950s, which led to its use in a drug marketed throughout eastern Europe. It is now an ingredient of the drug Reminyl, a treatment for Alzheimer's. Research continues into the uses of galantamine and other compounds derived from snowdrops; one protein is a proven insecticide and potentially has efficacy against HIV.

Interest in other species and in different forms began in the late 19th century, some, e.g., forms of *Galanthus plicatus*, arrived with returnees from the Crimean War of 1853–56. A little later E. A. Bowles did much to popularise the genus. In the 1980s, around 175 million bulbs were being exported from Turkey annually, but a complete ban on trading wild bulbs began in 1989, when *Galanthus* was added to the international conservation legislation CITES, and farmers in Turkey began to grow bulbs commercially.

Snowdrop growing is now a well-recognised gardening cult; in Britain, devotees (dubbed galanthophiles) organise "snowdrop lunches" to view them in January. Cultivars now number in the thousands, although these are usually little more than the natural variation of the wild species. In their natural range, a normal population would include variation; but further north, where reproduction is mostly clonal, this range is very limited, so anything a little different tends to be seen as special. Some cultivars change hands for considerable sums, and theft from gardens can be a problem.

Geranium Geraniaceae

The 260-odd species, all herbaceous, are very widely distributed. The majority are found in cool temperate regions of Eurasia; the rest are in the Americas and on various oceanic islands, including New Zealand. Members of the genus (whose name derives from the Greek for "crane") are commonly known as cranesbills. There are species from alpine and other harsh environments, but the majority are plants of open woodland, woodland edge, and grassland habitats, with relatively high levels of moisture and nutrients. Geraniums are overwhelmingly herbaceous in character, with basal leaves and flowering stems that carry smaller leaves.

Most are long-lived and clonal, but some, including quite a few annuals, are non-clonal. Clonality is expressed by constantly dividing rhizomes at, or just below, the soil surface, forming mats, or in some cases a more persistent crown with highly integrated shoots.

Geraniums tend to have a very flexible mode of growth, which enables them to penetrate dense vegetation and come out on top, or at least much nearer the light; in gardens, however, they tend to form dense clumps, which without support can collapse mid-season. Most species in cultivation have a strongly competitive character, but the pioneer tendencies of others show in prolific seeding. The species of warmer climes—the

Geranium sanguineum in what is a typical habitat for it—sand, in this case dunes on the South Wales coast.

Geranium collinum covering several kilometres of hillside in the Song Kul region of Kyrgyzstan, forming well over half the above-ground biomass of the vegetation.

Mediterranean basin and western Asia—tend to be annuals. There are some monocarpic or non-clonal short-lived perennials, too, e.g., *Geranium maderense* from Madeira. The ability of many species to grow at lower temperatures than many ornamental perennials has proved a boon to gardeners in northwestern Europe, with its long growing season, as the plants can compete with weeds during the winter.

Despite their being so widely distributed and common, little mythology or herbal practice is associated with geraniums, the exception being the North American *Geranium maculatum*, whose root (known as alum root) was used by both Native Americans and settlers for treating a wide range of conditions.

The central European species *Geranium phaeum*, *G. pratense*, *G. sanguineum*, and *G. sylvaticum* were occasionally cultivated in the early modern period and are illustrated in *Hortus Eystettensis* (1614). They were listed in early-20th-century books but featured little in nursery catalogues. The real interest in them can be said to have started with a pamphlet about them published in 1946 by English nurseryman Walter Ingwersen, with a number of other publications appearing from the 1950s onwards. Graham Stuart Thomas, appointed gardens advisor to Britain's National Trust in 1955, saw their value in low-maintenance ground cover plantings; they helped solve one of his problems: how to reduce maintenance in the vast acreage of country house gardens the organisation had just acquired from the nation's temporarily impoverished aristocracy.

Geranium macrorrhizum (left) and *G. phaeum*, drawn by Johann Carl Weber for *Die Alpenpflanzen Deutschlands und der Schweiz*, published in Munich in 1872. It was really only as wildflowers that most gardeners would have known *Geranium* species until the interest in them as garden plants grew from the 1950s onwards.

Geraniums have grown in popularity ever since, led by British gardeners, for the plants are ideal for the country's climate; they have not been so popular in more continental climates. New introductions from the wild and a wider distribution of species formerly found only in botanic gardens have joined an ever-widening number of hybrids. One of the most productive and inventive hybridisers has been Alan Bremner, a farmer in the remote Orkney Islands, off the north coast of Scotland, who makes his crosses purely as a hobby. A genus that was hardly hybridised a few decades ago becomes more complex by the year.

Gesneriads Gesneriaceae

The family Gesneriaceae has around 150 genera and 3,200 species, found spread across the tropics and subtropics, with a very few in temperate regions (e.g., southeastern Europe). The family includes three very familiar houseplant genera—*Sinningia* (gloxinia), *Saintpaulia* (African violet), and *Streptocarpus*—and several others cultivated in gardens. Gesneriads (as family members are known) tend to be non-woody perennials, although there are some woody ones, including small trees. Many are easily grown, compact, and have a vaguely cuddly demeanour, which has no doubt helped endear them to aficionados of indoor plants. For functional purposes, gesneriads can be understood by reference to their root structure, which has an impact on their cultivation requirements:

- Tuberous rooted. These are plants that have adapted to become dormant in areas with strongly seasonal rainfall patterns. They die back to a tuberous root, e.g, *Sinningia*.
- Rhizomatous. These similarly come from seasonally dry climates but survive as small scaly "tubers," which resemble tiny pine cones, e.g., *Achimenes*, *Kohleria*.
- Fibrous rooted. These very often originate from climates with more continuous rainfall patterns, although by no means always. Some are terrestrial or lithophytic, e.g., *Chirita*, *Episcia*, *Saintpaulia*, *Streptocarpus*, others epiphytes with semi-woody growth, e.g., *Aeschynanthus*, *Columnea*.

Among the non-woody species there is a very strong tendency to favour humid, shaded environments in forests in either montane tropical regions (especially cloud forest) or temperate regions with warm, humid summers. Many grow on rocks, damp cliff faces, banks, and streamsides, as well as forest floor situations where there is little competition. Those that become dry-season dormant are more likely to be from more open habitats. The epiphytes tend to have a branched semi-woody structure and hang down from their rooting position. Pollination is by insects, birds (especially for those with red or orange tubular flowers), and bats.

Of the most familiar genera in gardens, *Achimenes* and *Kohleria* are from Mexico southwards down the foothills of the Andes, and *Aeschynanthus* and *Columnea* are of

American and Southeast Asian origin, respectively. The two most commercially important genera are from Africa: *Saintpaulia* is found from near sea level to 2,200m, growing
in the humus of limestone and metamorphic rocks, always in shade, and *Streptocarpus*
is found in a wider range of habitats in the non-tropical areas of eastern and southern
Africa. The latter genus includes mono-
carpic species and semi-shrubby semi-
succulent species, as well as the familiar
clump-forming clonal perennials. Most have
to survive a dry season; many have survived
well in the open after deforestation. A few
have the extraordinary habit of producing
only one or two leaves during their entire
lives—the leaf grows continuously at the
base and wears away at the far end.

Saintpaulia ionantha* was brought to
the knowledge of Europeans around 1890,
through Walter von Saint Paul-Illaire, local
administrator of the German colony of Tan-
ganyika. He was an amateur botanist who
passed his finds onto European botanic
gardens. Suttons of England and Benary
of Germany made some attempt at developing the plant (in Germany, it was known
as Usambaraveilchen, after the area it was
found), but it was not until 1926 that it really
took off commercially, when the California
company of Armacost and Royston selected
varieties from seed obtained from these
companies and heavily promoted them as
"African violets." Nonstop breeding then
began in earnest—the first doubles appearing in 1939, the first pink in 1940—and by the
early 1960s a key problem, that of premature petal drop, was solved. The first African
violet show was held in 1946 in Atlanta—so
many people attended that the police were
called in to control the crowds.

The African violet became *the* flowering
plant for the modern centrally heated home
(at least until it was displaced by phalaenopsis orchids towards the end of the 20th cen-

JOHN INNES
1971

Streptocarpus was the subject and beneficiary of genetic research at
Britain's John Innes Institute for decades, and many commercially valuable varieties of the plant were the result. This is the cover of the Institute's 1971 annual report. John Innes Archives, courtesy of the John
Innes Foundation.

tury). The plants are at home at the temperature and humidity levels we like, and do not
need particularly high light levels. They are compact, and if the indoor gardener gets

truly carried away with them, veritable miniature borders can be created under artificial light. They bloom when young and can be propagated by the almost absurdly easy method of leaf cuttings (as can many gesneriads). Propagation also has a high tendency to produce somatic mutations, adding to the interest.

Sinningia speciosa was introduced from Brazil by Conrad Loddiges in 1817, as *Gloxinia speciosa*; the first hybrids began to appear in the 1830s, and during the second half of the century French, German, and Belgian growers systematically hybridised the plant. Species of *Achimenes* were hybridised extensively in the same period, many by Eduard Regel in Zurich; they remained fashionable into the 20th century, but interest

Achimenes were very popular 19th-century greenhouse plants, often displayed in hanging baskets; easily propagated and overwintered, they were soon embraced by anyone who could give them window space. From *Gardening Illustrated*, 1887.

has slowly declined. During their heyday they made a very rapid transition out of the greenhouses of the wealthy onto the windowsills of those who lived near the "big house," as the plants die down to tiny tubers in winter, and are very easy to start back into life again in the spring.

Cape primroses, as they used to be known (streps being a more familiar common name now), also made their entrance during the early 19th century. *Streptocarpus rexii* appeared at Kew in 1827, and both Messrs. Veitch and Lemoine were selling hybrids by 1862. Benary in Germany bred varieties with flowers to 10cm across during the early 20th century. The breeding behaviour attracted pioneer geneticists at Britain's John Innes Institute in the 1930s, with some of their varieties (e.g., 'Constant Nymph') entering commercial production. The plant was one of the first to be treated with x-ray radiation or chemical agents to increase the mutation rate; the genes of these "radiation mutants" then joined the modern commercial gene pool. Some 40 species were in cultivation by the mid-20th century, as well as a range of interspecific hybrids—indeed, what geneticists would call a hybrid swarm. Hybridisation is in fact frequent in nature and almost certainly has enabled new species to evolve.

This must be one of the first illustrations and attempts at selling African violets. From *Kelway's Manual* of 1894, the *Saintpaulia* is described as a "charming novelty which has created a sensation since its introduction." Courtesy of Kelways Plants Ltd.

All streps were blue and/or white until work at the John Innes Institute in the 1970s bred varieties with pink flowers. This sparked a revival of interest in *Streptocarpus*, with U.S. and Welsh breeders leading the way with an ever-widening range of colours and patterning. To some extent the genus has now replaced *Saintpaulia* as a popular flowering houseplant, the latter being alarmingly prone to turn to revolting mush when overwatered. The current wave of strep breeding favours clusters of many smaller flowers.

Gladiolus Iridaceae

The name is from the Latin for "sword," after the leaf shape. The approximately 260 species, all herbaceous plants that die down to a corm in the dormant season, are found from southern Europe down through Africa, with 250 being native to South Africa; it is from these South African species that the numerous garden hybrids descend. Gladiolus are typically found in open habitats, often grassland, with much variation between species in the preference for soil moisture. Many are very rare, found only in tiny areas, and are consequently very vulnerable to changes in land use or overgrazing. The flowers are pollinated by a range of bees, hawk moths, and sunbirds. Several species are used in traditional African herbal medicine, largely for gastrointestinal problems.

Southern European species have a long history of occasional cultivation, but with the introduction of the Cape species in the 17th century, interest in the genus took off. A steady stream of new introductions fed a process of continous and widespread hybridising, which began after 1820. The Ghent Group were bred in Belgium c. 1837, initially by the gardener to the Duke of Arenburg and later by van Houtte. British, Dutch, and French growers followed with more—the appeal was a strong palette of rich colours and ease of growth, as the plants could be readily dug up every autumn and stored over

Kelway's gladioli at a Royal Horticultural Society show in 1925, in the Society's hall in Westminster, London. Flower shows still happen here but far less frequently than in the past; during their heyday, they were fortnightly. Courtesy of Kelways Plants Ltd.

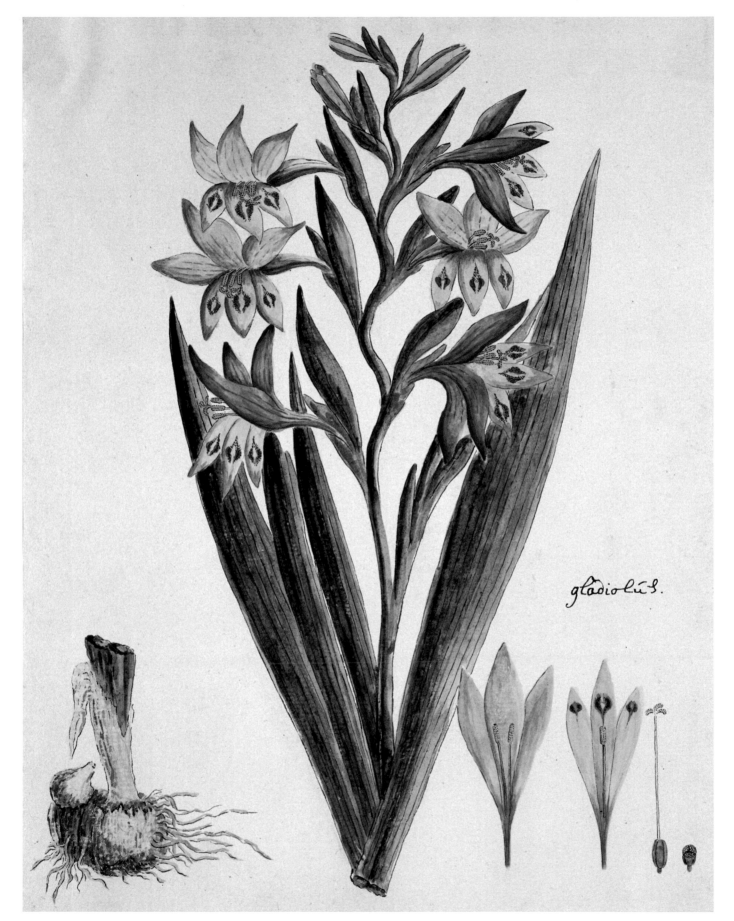

Gladiolus carneus. A drawing made during one of Robert Jacob Gordon's four expeditions to the Cape Province (South Africa), produced sometime between 1777 and 1786. Ink and chalk on paper. Gordon (1743–1795) was an explorer, artist, and naturalist of Scottish ancestry. Courtesy of the Rijksmuseum, Amsterdam.

winter as dry bulbs. From 1875, Lemoine produced a range of hybrids with contrasting markings on the petals, incorporating the latest discoveries from Natal and East Africa.

The c. 1890 addition of *Gladiolus dalenii* (as *G. primulinus*), introduced from the very edge of the Victoria Falls by the engineer who built a bridge across it, brought in yellow and orange shades, previously absent. The German William Pfitzer used this, and several other species not so far bred with, to produce a more delicately proportioned race—the Butterfly Hybrids. From 1890 onwards U.S. breeders took the flower up. Interest was probably at its peak in the period before World War II, with nurseries listing hundreds of cultivars. There was a big decline after the war, except in the communist block, where they remained popular and some new breeding was done. Their ease of growth as a crop for the short Russian summer was probably a reason for their popularity. Visitors to the "Socialist" countries would often be greeted at the airport by a woman in appropriate national costume clutching a huge bunch of the flowers.

OPPOSITE New Langprim Hybrids from the gladiolus supplement to Kelway's 1926 catalogue; these were the result of crossing large-flowered hybrids with *Gladiolus dalenii* (as *G. primulinus*) to create a more delicate flower, usually with shades of yellow. The Somerset nursery was one of the leaders in gladiolus breeding from the late 19th century onwards. Courtesy of Kelways Plants Ltd.

Grasses Poaceae

With more than 10,000 species, the grass family is the fifth-largest family of flowering plants and dominates many unforested habitats around the globe. The bamboos (Bambusoideae) are true grasses but considered separately; they retain some primitive characteristics but are not ancestral to other grasses. The supremely successful trick which the grasses developed early on (they date back to the Cretaceous) is that they make new growth from the base, not the stem tip, which means that they can survive grazing and fires. All are wind-pollinated, and their flowers are minimised to the point where it is difficult to see any relationship with other flowering plants. Physiologically, they generally need light to grow, and relatively high levels of phosphorus; on very poor soils, they tend to be replaced by subshrubs. Most grasses have the normal C_3 pathway for photosynthesis (these are the cool-season grasses); some however (including *Panicum* and *Miscanthus*) use the C_4 pathway: these warm-season grasses will start into growth only at higher temperatures, which can be a limitation at higher latitudes.

Grasses include both annual and perennial species, the latter, along with short-lived perennials (many *Bromus* spp., for example) are pioneers, while clonal perennial species are among the most effective of competitors; however, in many habitats, the most successful species are those which combine competitor, stress-tolerant, and pioneer strategies. Grasses typically dominate the middle stages of succession and may form relatively stable communities for some time if there is no nearby source of shrub and tree seed. While non-grass perennials may make an impact in the early years after clearing ground, grasses tend to dominate later on. The exceptions are places where the soil is relatively low in phosphorus, which reduces grass growth, allowing non-grasses to compete more effectively.

Most grasses are cespitose in habit, forming a tight clump, which after a certain point expands only very slowly, or indeed not at all, very often forming a tussocks. Cespitose grasses have root systems that are often very extensive and that tend to recycle

> How can these jewels of the garden have been virtually ignored for so long?
>
> (Foerster 1957)

nutrients from previous years' leaf litter—they effectively monopolise their immediate environment. Indeed, in tussocks the roots are largely ramifying the dead remains of previous years' growth. Some cespitose grasses, such as the North American *Sporobolus heterolepis*, can dominate vast areas. Non-cespitose grasses exist on a gradient, from those which slowly form mats (e.g., *Sesleria* spp.) to those which form tillers, runners which enable them to spread rapidly. Tillering grasses include those species of *Agrostis*, *Festuca*, and *Lolium* that give us pasture, meadow, and their civilised version—the lawn. Very rapidly tillering grasses include those of unstable environments (e.g., *Elymus* spp.), often found on sand dunes.

Grasses are inextricably linked with grazing animals and fire, both of which have, historically, kept vast areas of the globe free of forest; and humanity has, since very early on, greatly extended the world's grasslands, both through the use of fire and through the management of herds of cattle, sheep, etc. Those from northern Europe tend to think of grasses as typically forming turf, but these species are very much the minority globally. However—having been introduced throughout the world's temperate zones to provide forage for grazing animals (and of course lawns)—these turf-forming European grasses have spread and caused critical changes to many ecosystems. This is especially the case in North America, where the impact of these grasses on the landscape is rarely appreciated. The thick sward formed by turf grasses (in comparison to the gappy sward created by the hummocks of cespitose grasses) has massively reduced habitat for other species to seed and for small animals (including, for example, the chicks of ground-nesting birds).

Humanity has an intense and longstanding symbiotic relationship with grasses: we are utterly dependent on them for food (corn/maize, wheat, rice, etc.), and particular species have been singled out as being good for grazing or for handicraft use (although they have attracted surprisingly little mythology or folklore). Grasses have been invaluable sources of materials, notably for building (as in thatched roofs), basket making, and the provision of flooring (either as in the preindustrial practice of using loose grass [strewing] or in woven mats). Uses in medical herbalism have been minimal.

Grass has been cultivated as lawn in northern Europe since the Medieval period, but the need to cut it limited its use until the invention of the mechanical mower in the mid-19th century democratised the practice; previously teams of people with scythes and rakes would have been needed. With the European-led globalisation of the early modern period, the concept of the lawn was taken all over the world, more often than not to regions where European turf grasses either did not grow and had to be substituted with less effective alternatives, or which needed unsustainable quantities of water. In North America, the lawn became an integral part of a suburban lifestyle and a symbol of what it meant to live as an American; much has consequently been written about its cultural significance.

OPPOSITE "Otsuki Plain in Kai Province," from the series *Thirty-six Views of Mount Fuji* by Hiroshige; woodblock print produced in 1858. *Miscanthus* grasses with *Platycodon grandiflorus*, and what could be chrysanthemums and a patrinia. Courtesy of the Library of Congress.

Arundo donax (giant reed), planted outside the gates at a grand house in England, early 20th century. Illustration from the 1921 edition of William Robinson's *The English Flower Garden*. The plant is now rarely seen in gardens.

OPPOSITE *Avena glauca* (now *Helictotrichon parlatorei*) in a public planting in a housing estate in communist East Germany during the mid-1950s. From Foerster (1957), courtesy of Verlag Eugen Ulmer.

Four grasses, clockwise from right: *Briza maxima*, *Panicum virgatum*, *Molinia altissima* (now *M. caerulea* subsp. *arundinacea*), *Stipa pennata*, photographed by Karl Foerster (1957). The briza and stipa are minor plants in cultivation; the panicum and molinia are now among the most important ornamental grasses. Courtesy of Verlag Eugen Ulmer.

A growing interest in naturalistic planting from the late 19th century onwards has meant that turf grasses have come to been seen as a possible matrix for other plants, for example in wildflower meadows, or in Robinsonian wild garden plantings. North American gardeners first realised the ornamental potential of grass-dominated prairie habitats in the early 20th century, but the conformism of the suburban lawn reigned until the 1970s, when dissident voices began to speak up for native grassland as a viable alternative to an arguably unsustainable colonial import.

Grasses had a minor role in 19th- and early-20th-century ornamental horticulture: *Miscanthus* (then *Eulalia*) was recommended by Robinson as a specimen plant, and a number of annuals, such as *Briza maxima* and *Lagurus ovatus*, were grown, generally for showy seedheads. Interest in other grasses, the very similar sedges (*Carex* spp.), and wood rushes (*Luzula* spp.) grew slowly through the century; in many cases those grown were variegated varieties of rather aggressively spreading pasture or wetland species, such as gardener's garters (*Phalaris arundinacea*), the latter in particular giving all ornamental grasses a bad reputation for invasiveness. Pampas grass (*Cortaderia selloana*) had some popularity since being promoted by Robinson, but by the 1970s it was widely seen as a cliché and often inappropriately planted, particularly when landed in the middle of small suburban lawns.

The breakthrough for grasses came in 1957 with Karl Foerster's *Einzug der Gräser und Farne in die Gärten* (The advent of grasses and ferns in gardens). His opinions influenced two Germans who emigrated to America: Wolfgang Oehme (1930–2011) and Kurt Bluemel (1913–2014), the former a horticulturalist who teamed up with landscape designer James van Sweden to use grasses as a major element in a highly influential nature-inspired planting style, the latter a nurseryman. Back in Germany, leading plant breeder Ernst Pagels made many selections. In Britain Alan Bloom was the first to use them, in his nursery's display garden at Bressingham in eastern England. Whereas ornamental grasses had previously been grown largely for foliage interest, the new generation were being selected for structure and the combined effect of foliage and flower/seedheads.

Grasses took longer to take off in Britain, where the formulaic "tallest at the back, shortest at the front" style of border reduced their appeal; many grasses need backlighting to look their best. By the 1990s, however, several specialist nurseries were having an impact, with gardeners particularly appreciating the value of grasses in early winter. A more open planting style, associated with Dutch designer Piet Oudolf (b. 1944), has done much to raise their profile and commercial importance.

Smaller-growing coloured foliage grasses (and many *Carex* spp.) are especially popular in the garden centre trade, while larger species, which do not look good in pots, are more likely to be sold by mail order. Increasing numbers of North American species are being introduced into cultivation and good selections made—these find their way over the Atlantic relatively quickly.

Heathers Ericaceae

Calluna vulgaris, the only species in its genus, is correctly known as heather, while members of *Erica* are heaths. Contemporary usage tends to be to call them all heathers, along with the two species of *Daboecia*. *Calluna* (Greek for "to make beautiful," after its traditional use as a broom-making material) occupies the northern fringe of heatherdom (although with higher-altitude populations down to Turkey); *Daboecia* (named after St. Daboec of Wales) is found only on the Atlantic fringes of Europe; and *Erica* (from the Greek for *E. arborea*) is found across Europe, parts of North Africa and montane regions in East Africa, and the Cape of South Africa—where, in one of the most extraordinary bursts of evolutionary virtuosity in the natural world, 770 of its 860 species are found.

These are all subshrubs or occasionally shrubs, often forming a large proportion of the biomass where they live. Nearly all will grow only on acidic soils, although there are lime-tolerant exceptions such as *Erica carnea*, the ancestor of many popular winter-flowering garden cultivars. Heather habitat is typically open, exposed, and windy: Scottish moorland may approach a *Calluna vulgaris* monoculture, whereas Mediterranean maquis will have a richer array of species. Unlike many subshrubs, heathers regenerate well from the base, for example after a fire. Typically, all heathers and their associated mycorrhizal fungi thrive where soil chemistry reduces the growth rate of competitors, such as grasses.

Heather honey, made when hives are put out onto moorland in mid to late summer, when *Calluna vulgaris* flowers, is one of the finest European honeys. The tips of the plant are used to flavor beer, a traditional use which has been very much revived in the current fashion for microbreweries, and their dense twiggy growth is valued in making thatching, fencing, and items such as brooms. In Britain, Gypsies sold sprigs of heather for luck. Given its prevalence in Scotland (a legacy of massive deforestation), heather often serves as a symbol for the country.

Heathers entered cultivation in the mid-17th century; there is no mention of them in earlier herbals. The Cape species were extensively cultivated in early glasshouses from 1780 onwards. A notable first was a 30 × 15m parterre at England's Woburn Abbey, made around 1825. Later on in the century they were used as hardy bedding plants for winter decoration. Permanently planted heather gardens became popular in the late 19th century; Robinson had several at Gravetye, and they were promoted by Jekyll.

A. T. Johnson's *The Hardy Heaths* (1928) was a watershed in increasing their popularity. Named varieties were few until the late 1800s, but the rate of development gathered pace in the early 20th century. The first British nursery to specialise in them was Maxwell and Beale in Dorset in the 1920s; Douglas Fyfe Maxwell (1891–1963) had an eye for spotting natural variations and getting them into cultivation (see sidebar for his description of his wife finding what would become *Erica vagans* 'Mrs. D. F. Maxwell').

The Heather Society was formed in 1963 in Britain and the Gesellschaft der Heidefreunde in Germany in 1977. The 1960s and 1970s saw something of a boom in specialist heather nurseries in Britain, with interest also growing in the Netherlands, Germany, and the Pacific Northwest. English nurseryman Adrian Bloom brought them to the peak of their popularity during this time, combining them with conifers in island beds and promoting them for low-maintenance planting.

> After searching the downs [near Helston, Cornwall] for days, without finding any Heather of sufficient distinction to be worth while collecting, we had—at the time—turned our attention towards the absorbing if somewhat material occupation of gathering mushrooms, when my wife spotted a patch of Heather of an unusual colour on the field bank.
>
> (Maxwell 1927)

Hebe Scrophulariaceae

Named after the Greek goddess of youth, the genus includes around 100 species, 80 of them in New Zealand, the remainder elsewhere in Australasia and the southern tip of South America. Originally classified in *Veronica*, the name *Hebe* was initially suggested for the antipodean woody species in 1789, embraced by the New Zealand Institute in 1926, and finally accepted by London's Royal Horticultural Society in 1985. *Parahebe* is a closely related genus with a more weakly developed subshrub type of growth and a prostrate habit.

Hebe species are found all across New Zealand, their shape and hardiness very much reflecting the latitude and altitude of their nativity. Those from the warmer north have larger leaves; those from the mountains in the south tend to be compact, with smaller leaves. The "whipcord" species, where the leaves are reduced to scale-like structures, are typically moorland or alpine species. A notable features of hebes is their wind tolerance. Many species, however, are not tolerant of cold temperatures. Hebes are all subshrubs, with only limited powers of basal regeneration and a similarly limited lifespan, therefore, with many deteriorating markedly after around 10 years; smaller species can live for decades but tend to lose their shape.

Plants were first introduced to Britain in 1876, with a constant trickle since. Some cultivars are selections of distinctive wild plants; many others are hybrids, the result of spontaneous seeding in British gardens. A Hebe Society was set up in London in 1985. Too short-lived to be useful as major elements in planting design, their hummocky shapes, evergreen nature, and ease of propagation have nevertheless made them popular in the landscape industry. They are occasionally used in bedding schemes and, in parks in South America, even parterres.

Hedera Araliaceae

Hedera (from the Latin for the plant) contains around 15 species. All ivies, as they are known, are evergreen woody climbers or ground creepers, from Eurasia and North Africa. They are distinctive for the aerial roots, which cling to tree trunks and buildings through a complex process, combining both physical and chemical ways of bonding. The plants climb until they can climb no more or they reach a certain maximum height, after which they become arborescent—no longer climbing, woodier, and with no aerial roots. Changes in leaf shape during the plant's life is characteristic of Araliaceae; in the arborescent phase, ivy leaves develop a simpler shape. Ivies are generally to be found, and successfully grown, in climates that do not have severe winters or drought. They play an important ecological role in woodland, notably for providing generous nectar for pollinators and habitat for invertebrates and birds; in North America, however, ivies are potentially very damaging to forest floor plant communities, as they can form a suffocating monoculture.

The damage that ivy supposedly does to trees and buildings is the subject of much discussion. The consensus is that it impacts negatively only on older and declining trees, which are losing their canopy, so allowing light to reach the ivy; an exception is that healthy conifers with an open canopy can be stressed by ivy growth, which adds to wind rock. On buildings, good maintenance is certainly required to limit the growth of ivy, but it can add appreciably to insulation.

Traditionally associated with Bacchus, the Roman god of wine, ivy was long used to advertise inns. There was also something of an association with death in European cultures, and being evergreen gets it noticed at Christmas. Historically there has been some medical use, and it is occasionally used as emergency animal fodder, although it is toxic in large doses. The plant has an undeserved reputation for the degree of its toxicity, partly because of confusion with the North American poison ivy, a totally unrelated and clearly very different plant, even to the extent that British schoolchildren refer to *Hedera helix* (English ivy) as "poison ivy"—an example perhaps of the over-powering strength of American popular culture.

[A]nd monstrous ivy-stems
Claspt the gray walls with hairy-
    fibred arms,
And suck'd the joining of the
    stones, and look'd
A knot, beneath, of snakes,
    aloft, a grove.

(Alfred Lord Tennyson 1809–1892)

Evergreen, easy to grow, and flexible, ivy was a decorative boon to our ancestors, its uses shaped over time by prevailing garden fashion. During the great era of the formal garden, from the 17th to the early 18th century, ivy was trained and clipped this way and that, as standards, and over all sorts of frameworks; and then with the advent of the naturalistic "English landscape," it became invaluable for letting rip on ruins to make them more romantic and "picturesque"—no grotto or hermitage was complete without it. Ivy's heyday was the Victorian era, its colour appealing to the dour aesthetic of the time, and its ability to cope with air pollution a definite plus. Numerous books suggested how it could be used as a houseplant, trained over bowers in the drawing room, up pillars, around fireplaces, etc. Vast numbers of potted plants trained up columns or over trellises were on sale in the markets, enabling almost instant effects to be created, inside or out. A particular Victorian favourite, and one perhaps worth reviving, is the "fedge," where ivy is grown up a fence—I have seen this done over a substantial metal gate, so making a fedge that can be swung open.

A variegated ivy—which according to Pliny the Elder was carried as decoration in a victory parade by Alexander's army in the 4th century BC—is the first known record of a variegated plant in the West. Ivies were taken to North America in the 18th century and became popular, especially to give an air of antiquity to buildings, which would hint at Old World respectability (as in "Ivy League" universities). Other species (e.g., *Hedera colchica* and *H. canariensis*) were introduced in the 19th century. Variable at the best of times, variegated sports or forms with novel leaf forms were selected and propagated, so that Shirley Hibberd (1872) could record 200 in his monograph on the genus. More recently, the realisation that the plant can become invasive has limited its popularity: Portland, Oregon, even has its own No Ivy League (est. 1994).

Helianthus Asteraceae

Helianthus (Greek for "sun" and "flower") is an unusual genus in that it includes a major agricultural crop, the sunflower (*H. annuus*), as well as ornamentals. All species are herbaceous and overwhelmingly North American (only three of the 51 species are South American); all are plants of open spaces or woodland edge habitats, generally on fertile and moist soils. Apart from *H. annuus*, they are long-lived competitive perennials, many of which have a strongly running guerrilla root system—this is more about finding new territory than conquest, but it has made gardeners wary of them.

The annual sunflower (*Helianthus annuus*) is both an important crop plant (for forage and oil) and a popular ornamental. First cultivated in Central America, it was one of the first plants introduced to the Old World from the New. The seeds were an important protein source for Native Americans, while any species with tuberous roots was eaten as a carbohydrate—*H. tuberosus* (Jerusalem artichoke) being taken into cultivation as a vegetable in the 1600s. A recent use of these plants is in phytoremediation, as plants can concentrate elemental pollutants in their tissue; *H. annuus* was used to help clean up after the 1986 Chernobyl disaster.

Graham Stuart Thomas grumbles, 'I cannot write about these with any enthusiasm', and proceeds to summarize a whole array of these rampant plants.

(Gerritsen and Oudolf 2000)

Cultivated since prehistoric times by Native Americans and introduced first to Spain in 1568, the sunflower spread rapidly to the rest of Europe, astonishing people with its height and high level of seed production. Variations in flowering habit (one big or lots of smaller heads) were soon noted, and different strains developed. Its use in commercial oil production seems to have begun in 18th-century Prussia, incidentally also an early centre for sugar beet development and production. Sunflowers are popular children's plants because they can grow so tall; giant varieties are a variant of the agricultural varieties where one large head grows at the top of a tall stem. The cut flower trade and most ornamental planting needs something different—those with multiple smaller heads or varieties with red-brown or double flowers.

Helianthus ×multiflorus (a hybrid of *H. annuus* and *H. decapetalus*) possibly originated in Spain in the 16th century. Various other, perennial, species, including *H. giganteus*, were introduced from the 17th century on; this constant trickle of species has given gardeners a succession of strong late-season yellows. In particular, a number of ornamental varieties under the names *H. ×multiflorus* and *H. decapetalus* have been grown (and their names thoroughly confused); their popularity is based on uncompromising yellow and ease of growth, which has generally meant a rapid translation from botanic garden to country cottage. 'Lemon Queen', a late-20th-century variety of obscure origin, is unusual in having pale flowers and a non-running habit; it remains one of the most successful perennial sunflowers.

Helianthus mollis and *Solidago altissima* in a man-made meadow, southern Appalachians.

Helleborus Ranunculaceae

Sixteen perennial species are found from southern China across to northwestern Europe, with a centre of diversity in southeastern Europe. It is difficult to define species, and levels of natural variation are high. The name is from the Greek for another poisonous plant, probably *Veratrum*. Two growth-form categories are recognised. Caulescent species (e.g., *Helleborus foetidus*) develop a semi-woody stem, several being usually produced annually; these tend to be non-clonal and can be short-lived plants. Acaulescent species have leaves that spring up directly from the crown of the plant; they are clonal and long-lived but spread very slowly. The most prominent feature of the flowers are the sepals, which function as petals; the petals themselves evolved into nectaries, special structures at the base of the sepals which hold the nectar.

Hellebores are plants of mature woodland, wherever light levels allow, and woodland edge habitats. As many gardeners realise, they seed profusely; in the wild this is probably a means by which a rare event creating disturbance and increased light at ground level, such as a tree falling, is then exploited—with seed germinating en masse, leading to a new population, which may then survive for decades beneath younger trees. The plants contain a number of ranunculins, irritant and toxic chemicals that discourage animal predation and possibly infection by microorganisms.

Hellebores were used as an oracle flower in Medieval Europe, with 12 buds being put into water on Christmas night, each one representing a month, so that the next morning the open buds would predict good weather, the closed bad. The species best known to our ancestors, *Helleborus niger* (Christmas rose), was used in early modern Europe as a treatment for a variety of conditions; overdosing however could be lethal. It had many mythological associations, particularly with the curing of mental illness. Greek collectors were said to have not dug the plant up without first ritually circling it with a sword, while reciting prayers to the appropriate gods.

According to German botanist Otto Brunfels (1488–1534), hellebores were even then well known in gardens. In the mid-19th century, breeders in Germany began to create hybrids, largely between different forms of the very variable *Helleborus orientalis*. Nursery catalogues of the time list almost all the coloured and spotted variants available today. Further genetic variation was added when plants from the Caucasus came to Germany, via the St. Petersburg Botanical Garden. The first forms of *H. orientalis* came to Britain at around the same time, when a number of amateur breeders exchanged plants and seeds, creating around 50 selections by the 1880s. In 1890 Victor Schiffner of Prague University published the first book on the genus, *Monographia Hellebororum*.

Interest in breeding slumped during the 1920s, and much was lost in World War II; *Helleborus niger* survived best, becoming the most commonly available species for some decades afterwards. One postwar English grower became obsessed with hellebores, however: Helen Ballard (1908–1995) taught herself German so that she could research the genus. Her breeding very much revived the plant in Britain. In her old age she sent some of her plants to Germany, where Gisela Schmiemann continued

Ashwood Nurseries in the English Midlands is one of the global leaders in hellebore breeding. This is their stock plant nursery, where painstaking hand-pollination every spring is used to continually improve the seed strains that are used to produce their plants.

her work; seed was then distributed to Japan, Australia, New Zealand, and the United States. Hellebores were introduced to the United States in the 18th century, but breeding did not begin there until the mid-20th century; interest in the plants has since grown in concert with the rise of interest in them in Europe.

Several German breeders, notably Heinz Klose and Günther Jürgl, produced the first doubles in the 1980s. In England, Eric Smith and Jim Archibald worked on them in the 1960s, while Elizabeth Strangman played a key role in developing the modern range of hybrids. Today, John Massey's range has a particularly high profile; he prefers to develop seed strains with a higher level of consistency than named cultivars—this makes sense, as vegetative propagation is painfully slow and expensive.

Hellebores wilt notoriously rapidly when cut but keep well when cut close to the flower base and floated in water. This method is even used when exhibiting the flowers at competitive shows.

Hemerocallis Hemerocallidaceae (Liliaceae)

The name is from the Greek for "day" and "beauty"—a distinguishing mark of the genus is that the flowers open for only a day (hence, daylilies). The 18 species are found across Eurasia, with most in the Far East. The relationship of the Hemerocallidaceae to other formerly "lily family" plants has been much disputed; current thinking puts *Hemerocallis* in its own family, although some botanists would put it in Xanthorrhoeaceae, most of whose members are southern hemisphere. *Phormium* is regarded as being particuarly close. *Hemerocallis* species can be found in a wide range of habitats, including mountain meadow and coastal situations; the common factor is sun or light shade, with moderately high levels of moisture and fertility. These are clonal perennials, forming dense, competitive, persistent clumps and often surviving in abandoned gardens; some (e.g., *H. fulva*) have more strongly running underground rhizomes. Although they are cold hardy, daylilies thrive particularly well in climates with hot, humid summers. They are listed as potentially invasive in some U.S. states. Hybridisation has resulted in a plethora of cultivars, which are divided into a variety of sections, based largely on flower form—trumpet, flat, flaring, star, spider, ruffled, etc.

Daylilies were an important source of food and medicine in China; the first book about them was published there in 304 BCE. Their medicinal uses were various, including for treating cases of poisoning. The flower buds are commonly eaten in Chinese cuisine, usually being prepared from dried buds, there being a popular belief that fresh buds are harmful. Young shoots and the tuberous roots may also be eaten, and evidence suggests that hemerocallis extracts may be useful in treating depression.

Plants were introduced into cultivation from the 16th century onwards; the first hybridisation was carried out by the English schoolteacher George Yeld (1845–1938), with 'Apricot' in 1892, followed by nurseryman Amos Perry. The real father of this most obsessively hybridised plant, however, was the American Arlow B. Stout (1876–1957). Stout was a geneticist and plant breeder, working with everything from poplars to potatoes. Daylily breeding was a hobby for him—he was renowned for leaving his work at the New York Botanical Garden whenever possible and immersing himself in his private plot of the plants. His greatest achievement was to use *Hemerocallis fulva* var. *rosea* in a systematic breeding program; yellow is the usual daylily colour, but rosea's genes introduced a new spectrum of pinks and reds. In the end, his obsession led to the introduction of 83 high-quality hybrids.

Hemerocallis is ideal for amateur breeding—easy to grow, remarkably easy to cross, and largely pest- and disease-resistant. Practically any seedling from a cross looks good, and like proud parents, it seems that many breeders have been unable to rate one

> Daylilies are dangerous, however, and absolutely addicting, except for the occasional gardener with a will of steel and a heart of carborundum. Someone gave me one once and I was trapped. . . . The addiction may go into remission, but it never goes away.
>
> (Allen Lacy 1935–2015)

Hemerocallis littorea growing on the clifftops of the Uzu Peninsula, south of Tokyo, in very windswept *Chamaecyparis obtusa* forest alongside *Ajania pacifica* (silvery foliage) and *Farfugium japonicum.*

of their children any higher than another, let alone to judge them objectively against others. A vast array of cultivars has been the result—some 75,000 registered with the American Hemerocallis Society (est. 1946) by early 2013, at a rate of around 3,000 a year. The society's membership of over 10,000 forms a vibrant and lively community and, perhaps more than any other specialist plant society, involves its members in breeding, including laboratory-based techniques. Treatment with colchicine induces chromosome doubling; even one use of colchicine converts old cultivars into tetraploids with stronger stems, better and more robust flower shapes, and petals that have an extraordinary sense of visual depth, allowing for the expression of very intense colours.

A bunch of flowers—including a tulip, an auricula, a convolvulus, and roses—topped by a daylily; watercolour on paper, by Henriëtte Geertruida Knip (1793–1842), a well-known Dutch flower painter. Courtesy of the Rijksmuseum, Amsterdam.

Heuchera Saxifragaceae

The 37 species of *Heuchera*, perennials all, are found throughout North America but have a centre of diversity in the Pacific Northwest. They are thought to have originated in the Eocene, when the disappearance of the Bering Land Bridge left them isolated from the rest of the Saxifragaceae on the American side of the Pacific. The eastern North American species are forest floor dwellers, the western are found both on the forest floor and on wet cliffs or in rock crevices; some western species are also notably very tolerant of drought. Plant growth is clonal, clump-forming, and persistent, with almost woody rhizomes. Many cultivars, however, have acquired a reputation for being short-lived; possibly this is because in cultivation they are deprived of the thick layer of leaf litter that normally covers the rhizomes in nature, allowing them to root in and supporting continual growth. There is a tendency to be semi-evergreen. *Tiarella* is very close; hybrids between the two genera are classed as ×*Heucherella*. The name honors Heinrich von Heucher, an 18th-century Austrian physician.

Several species were used by Native Americans to treat a wide range of ailments. There is confusion on this point, however: the plants are commonly known as alum root, yet this is also the name for *Geranium maculatum*, which has been of far greater use in herbalism.

Heuchera micrantha in the Columbia River Gorge, Oregon, on a wet rockface, a typical habitat for the species.

Heuchera sanguinea hybrid from Vaughan's 1936 catalogue.

Heuchera americana came to England via Tradescant in 1656. It appeared in German gardens in the late 17th century, but for a long time was grown primarily in glasshouses or orangeries for its winter foliage interest. Lemoine crossed it with *H. sanguinea*, which had been introduced to England in 1880, to produce a range of hybrids, grown mostly for their floral rather than foliage interest. Alan Bloom carried out some breeding, and in 1934 put on a display composed entirely of heucheras at the Chelsea Flower Show; by the 1950s he had a good seed strain, Bressingham Hybrids, and was making intergeneric crosses with *Tiarella*. Many American breeders have worked on the genus, notably Charles Oliver and Dan Heims, producing a vast array of cultivars with an extensive array of foliage colours. Hybrids so far are derived primarily from *Heuchera americana*, *H. micrantha* (valued for its ruffle-edged leaves), the coastal dweller *H. maxima*, drought-tolerant *H. sanguinea*, and *H. villosa*.

Hibiscus Malvaceae

There are some 750 species of *Hibiscus*, whose name is derived from the Greek for the closely related mallow. Overwhelmingly they are found in the world's tropical regions, both Old and New Worlds, and include trees, shrubs, perennials, and annuals. The diversity of the genus reflects perhaps that it has become something of a botanical ragbag, and a scientific paper threatened that "upheavals concerning *Hibiscus*, one of the world's most popular horticultural plant genera, will be difficult to avoid" (Pfeil et al. 2002). Species are generally plants of well-watered and fertile soils, but generalisations across such a varied genus are difficult to make.

Hibiscus syriacus is a gold mine for a breeder because of the potential for myriad flower forms quickly and fully expressing their attributes.

(Dirr 1998)

Since the 7th century at least, *Hibiscus syriacus* has been grown in Korea, where (in the South) it is the national flower, seen as embodying the tenacity and survival instincts of the Korean people and their national culture (hence its Korean name, *mugunghwa*, "everlasting flower"). During the Silla period (57 BC to 935 AD) the country was known as "hibiscus land." A stylised version of the flower's five petals is a common part of Korean national iconography. The flowers are eaten in China and Korea, and tea can be made from the leaves. More important gastronomically is *H. sabdariffa* (roselle), which is used as a vegetable in many tropical countries and is even more popular as the source of a sourly refreshing red tea; this has been shown to have positive benefits in treating high blood pressure but is so acid it can damage tooth enamel. Some species are used for making paper, e.g., *H. cannabinus* (kenaf).

Hibiscus syriacus was introduced to Germany from present-day Lebanon in the 16th century, although there had been a considerably longer history of cultivation in the Middle East, and longer still in eastern Asia. Colour variants were often mentioned by early

European garden writers, who knew it only as an orangery or greenhouse plant. Doubles were known by the early 19th century. It has been very successful over much of the United States, where it is known as rose of Sharon (confusingly, as in Britain this common name is applied to *Hypericum calycinum*). In warmer climates, the plants can seed problematically—a challenge taken up by Don Egolf, of the U.S. National Arboretum Plant Introduction Program, who since 1970 has released a series of lavishly flowering hybrids that are triploid and therefore sterile.

In those temperate regions with warm and humid summers, the range in cultivation is boosted by a number of herbaceous perennial species from the U.S. Southeast, which have enormous flowers, e.g., *Hibiscus moscheutos*. In the tropics, *H. rosa-sinensis* is the most popular species in cultivation, with many varieties in a wide range of colours. A native of eastern Asia it has been in cultivation for many centuries—it is not clear how many. It is a polyploid, which adds much complexity and indeed unpredictability to its genetics; it is a very popular plant for both commercial and amateur breeding across the tropics, with many local societies promoting the plant and encouraging breeding.

A hybrid involving *Hibiscus moscheutos* from Vaughan's 1936 catalogue.

Hosta Asparagaceae

The genus, whose name honors Austrian botanist Nicholas Thomas Host (1771–1834), comprises some 45 species from eastern Asia. Hostas are found in a wide range of moist habitats, from wet cliff faces to mountain meadows and woodland, mostly where the soil is freely drained; only a few are ever found in waterlogged soil (i.e., true marshland conditions), and many are found in deep shade. All are long-lived herbaceous perennials, with a competitive character and a spreading habit, which varies from steadily clump-forming to guerrilla. Like its close relative, *Hemerocallis*, the genus is of great use in smaller scale commercial landscaping as well as having a devoted cult following.

Hostas are edible if somewhat tasteless, and in some parts of Japan they are regularly eaten as a vegetable. It has been suggested they'd make a very productive forest garden crop, given how prolific they are. The emerging rolled-up leaf is wonderfully tender; mature leaves may be treated like spinach.

Hostas began to appear in art in the Edo period in Japan. Some selection and hybridising took place during this time, but the plants were not of major interest. Kaempfer and Thunberg both obtained plants, and von Siebold was able to bring plants back to Europe in 1829. With active

OPPOSITE A Japanese print showing a flower arrangement with some hosta foliage at the base, along with irises, lilies, daylilies, what could be a *Spiraea* sp., and (rather improbably) a pine bonsai. Produced 1765–66, in the manner of Ippitsusai Buncho. Courtesy of the Rijksmuseum, Amsterdam.

trade between Japan and the West during the Meiji era, an increasing number were exported from the nurseries in Yokohama, who specialised in growing for foreign markets, including by an American working in the country, Thomas Hogg, whose brother James ran a nursery in the United States. Western trade stimulated greater Japanese interest, with growers systematically collecting and naming plants to sell to Westerners, especially if they were variegated. Western interest in hostas, in part as shade plants, established quickly, with particularly widespread use in U.S. parks and cemeteries; their tropical appearance was much appreciated at a time when bedded-out exotica was very fashionable. Georg Arends offered his first cross in 1905; like other early breeders, he focussed on flowers and flowering time as much as for foliage. Foerster also bred several; he liked to juxtapose large leafy plants with grasses and ferns in a combination he referred to as *Dreiklang*—the sounding of three notes in perfect harmony.

From such promising early beginnings, interest remained rather static until the 1960s—when nurseries in the United States, United Kingdom, and the Netherlands began to specialise in the plants; and enthusiasts began to make trips to Japan to look for new species and variants. Ensuing administrative details were handled by the American Hosta Society (est. 1968) and the International Registration Authority for hosta names (est. 1969). A key person in the revival and breeding of hostas was Eric Smith of The Plantsman Nursery in southern England, who listed 21 in 1970.

Post-1960s developments have considerably widened the gene pool, with hundreds of cultivars now available. These new introductions feature improved leaf substance, pleating, and of course endless permutations in variegation; in contrast to earlier times, interest in the flowers has been minimal. Hostas are particularly popular in the United States, where the hot humid summers over much of the country are very similar to those experienced in the plants' homeland; here breeders and enthusiasts have built up a lively hosta-growing cult.

Hosta sp. growing out of a water seepage in rocks, Ibukiyama, Shiga Prefecture, Japan.

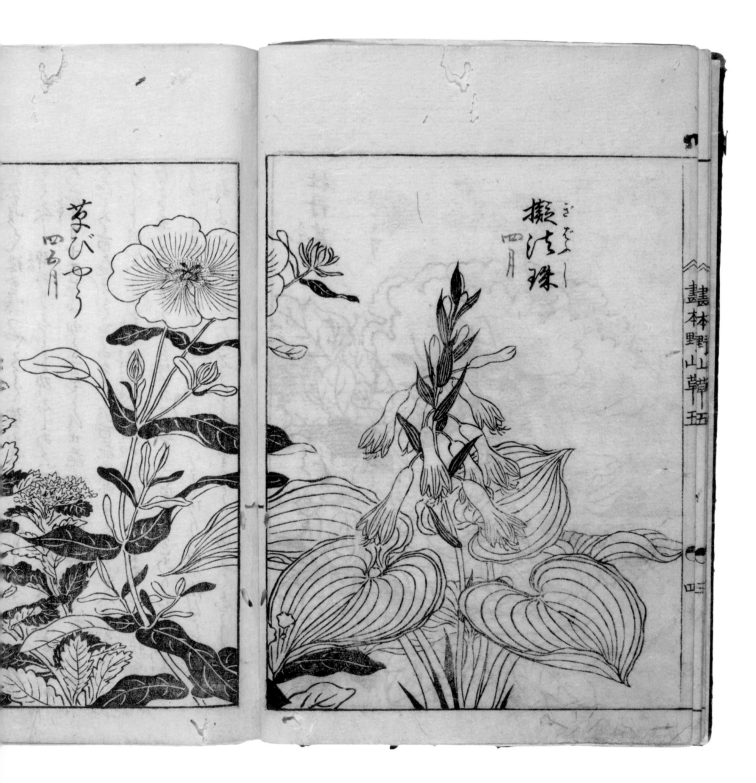

A *Hosta* sp. dominates this spread from a volume compiled by Tachibana Tamotsukuni, 1755. The far left is a *Spiraea* sp., the other plant is unidentified. Courtesy of Chiba University Library.

Hyacinthus Asparagaceae

Classical legend has it that the flower sprang from the spilled blood of the Spartan prince Hyacinthus, accidentally killed by his lover Apollo during a discus-throwing contest. There are three species, all geophytes, from the eastern Mediterranean and Middle East. Like many bulbs from the Mediterranean region, wild hyacinths are found in open or woodland edge habitats—wherever moisture and light are sufficient for growth during winter and early spring. Dormancy begins in early summer as temperatures rise.

The main use of the plants in premodern times was as a source of starch for stiffening the ruffs of men's clothing, as well as for making glue, which had a particular application in bookbinding.

The first records of *Hyacinthus orientalis* in cultivation are from the Arab world, but after the Ottoman takeover of the Middle East from the 15th century onwards, it was grown more extensively. It was probably introduced to Europe through the botanic garden of Padua, in Italy, in the 16th century. Becoming widespread quickly, it was regarded as plentiful by the early 17th century, popular for both its visual

Hyacinthus orientalis, growing alongside *Ranunculus kochii* and *Gagea bohemica* above Kemaliye, Turkey.

appeal and its fragrance. *Hortus Eystettensis* (1614) shows 17 varieties. The first book devoted entirely to them was published in Halle (Germany) in 1665. By 1725, Miller claimed that the Dutch were growing some 2,000 varieties, more sober German commentators that there were "only" 240 double and 100 single varieties. As with tulips, prices reached dizzying heights, plants in the 1760s going for sums equivalent to over £10,000 per bulb, at today's prices.

The English imported considerable quantities from the Dutch from the early 18th century until the mid-19th century, when home suppliers began to produce more. They were very popular in Germany where millions of bulbs were grown and sold in the early 19th century, including in the vicinity of Berlin; industrialisation brought about a decline in home production, which Dutch exporters were only too ready to make up.

A distinct feature of the cultivated hyacinth has been its tendency to produce dramatic colour breaks. In the 19th century there were strong national preferences: the British preferred pink and white, the Russians blue, and the Germans a mixture but with large flowers. Towards the end of the century, German breeders such as Benary began to produce dark-toned and subtle transitional colours, possibly reflecting the wide variety of colours which the new chemical dye industry was able to produce for textiles. Since World War II, the Netherlands has been the centre of breeding and production.

Hydrangea Hydrangeaceae

Hydrangea includes some 35 species of small trees, shrubs, or shrubby climbers, found in eastern and southeastern Asia and the Americas. The name comes from the Greek for a water vessel, after the shape of the fruit. Hydrangeas are plants of regions with warm and humid summer climates, the shrubby kinds growing typically in woodland edge habitats, and in Japan, along the coast. The climbing species all use aerial roots to hang onto the trunks and branches of host trees. Some shrubby species sucker to form extensive colonies. The genus is divided into two sections, 11 species being in *Hydrangea*, all deciduous shrubs and climbers (e.g., *H. anomala* subsp. *petiolaris*), the remainder in *Cornidia*—evergreen climbers with a disjunct distribution in Mexico and the Andes, of which very few are in cultivation. The two most important species in cultivation are *H. macrophylla*, which has both lacecap and mophead varieties, and *H. serrata*, all lacecap.

Dewdrops on the four sepals of the
   garden hydrangea,
glistening softly in the evening moon
   of the fading summer,
quite impossible to ignore.

(Toshinari Fujiwara, 12th-century Japanese poet, quoted in Kashioka and Ogisu 1997)

For gardeners, the distinction between lacecap and mophead is crucial. Plants are naturally lacecaps, looking superficially like many *Viburnum* species (which are unrelated) or indeed herbaceous plants of the Apiaceae, with a large number of small fertile florets being surrounded by a corona of larger sterile ones, the latter attracting

Hydrangea serrata growing in woodland edge in Hokkaido, Japan, showing the wide range of flower colour typical of the species in the wild.

pollinators. Mutation may result in nearly all the fertile florets being replaced by sterile ones—a turn of events that renders the flowers totally dysfunctional as far as wild plants are concerned but which led to great interest from humanity. Besides being treasured as ornamentals, hydrangeas have been used in both eastern Asian and Native American herbal traditions, for treating various ailments, with a particular reputation for treating kidney stones. Although the plant is toxic, a sweet-tasting tea is made from the leaves in Japan and Korea.

The first mention of hydrangeas in any text was in 8th-century Japanese poetry, with it being noted that the flowers have different colours in different places; doubles were noted too. Wild forms with mopheads are known from various places in the country, and it is likely these were first taken into cultivation during the Heien period. Over time the plants became more popular but were never a major subject of interest: possibly because wild plants were so common and easy to grow, they lacked any real cachet. The irony is that it was only after World War II, with introductions to Japan of plants from the United States, that hydrangeas have become really widespread in Japan.

The first hydrangea to be introduced to Europe was the American *Hydrangea arborescens*, which John Bartram sent to England in 1736. Thereafter, confusion reigned. Plants coming from Japan were often the cultivated forms with all-sterile heads, lacking the floral parts which enabled botanists to classify plants. Kaempfer thought the plants he was seeing in Japan were an elder, Thunberg that his were viburnums. Further confusion was added in 1792 when an English botanist realised that these plants were hydrangeas and classified them as *H. hortensis*—"of gardens"; the French meanwhile had coincidentally called the plant *Hortensia opuloides*, after Hortense de Nassau, the daughter of a socially prominent botanist.

Sir Joseph Banks is credited with bringing the first mophead, now regarded as *Hydrangea macrophylla*, to Kew, in 1796. Further introductions brought propagating material to the West; their almost over-the-top blowsiness appealed, and they were propagated and sold by the thousands, in France and Britain, often being forced. In Germany it was not just the flowers which appealed but the foliage, as the plants would leaf up early indoors, a welcome sight at the end of a long winter. A German priest, writing in 1818, recommended including iron ochre in the potting soil, as this would often turn flowers blue. He seems to have been the first to realise that some varieties flowered pink or blue, depending on the soil chemistry: more acid conditions generally allow enough soluble iron to be available for the blue reaction; more alkalinity, and the iron stays insoluble and the flowers are pink. Breeding, however, was hampered by the fact that mopheads were sterile.

Veitch employees collected several forms, including lacecaps, in Japan in 1877–79 and, not realising their commercial value, passed some on to colleagues in France. Lemoine et fils and Louis Cayeux were among the nurseries who saw the potential. The first few years of the 20th century saw Lemoine produce several lacecap hybrids and some mopheads, as the enterprising Lemoine family had examined the mopheads plants carefully and found occasional fertile flowers, enough to breed with. Their flowering in summer and tolerance of shade made them very popular in central and southern

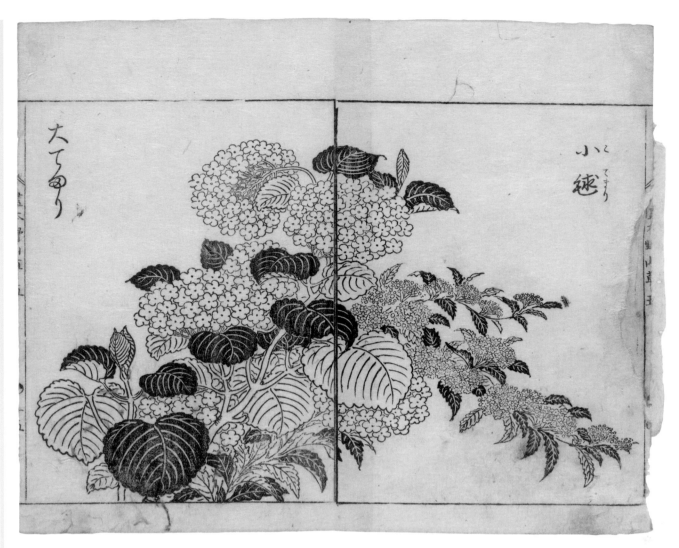

A mophead hydrangea in a Japanese book of 1755, *Illustrated Book of Mountain Plants*, by Yasukuni Tachibana (1715–1792). Courtesy of the Library of Congress.

Europe, while smaller-growing ones were selected for pot culture. They took off particularly in areas on the Atlantic fringes of Europe, where a combination of mild, humid climates and acid soils proved ideal, not just for cultivation but for true blue flowers.

E. H. Wilson found wild plants on a Japanese seashore in 1917 and realised that they were the wild ancestors of the Japanese garden plants, the origin of which up until then had been a mystery. This period saw the introduction of many other species into cultivation, some of which, such as *Hydrangea aspera*, have been grown for the quality of their foliage rather than their flowers. Hydrangea breeding continued strong throughout the 20th century; the most significant recent developments have perhaps been an emphasis on the improvement of the North American native *H. quercifolia*.

Ilex Aquifoliaceae

There are around 400 species of holly: trees, shrubs, and woody climbers. The genus (named after the Greek for the evergreen holm oak, *Quercus ilex*) dates back to the Paleocene; it diversified before the continents parted, and then died out in many places, leading to a highly disjunct distribution—there are many species in the Americas, particularly the tropics, and in Southeast Asia down to New Guinea, but only one in Africa and one in Europe. In New Zealand, hollies are known only as fossils. Hollies tend to be stress-tolerant plants, the fact that most are evergreen making them effective at making the most of life as understorey beneath taller species. The flowers are very attractive to pollinators, often enticing them away from showier flowers in the vicinity. The fruit tends to be poor quality and consequently very persistent, which makes them useful for gardeners and as famine-food for birds at the very end of the winter.

Ilex aquifolium has been extensively used across Eurasia and the Middle East since ancient times for decoration at winter festivals; the Romans used it to celebrate Saturnalia. There are major links with pagan beliefs in Europe, but usually these are given a Christian veneer. It has been much used in folk-divination and healing rituals, which perhaps is behind the very wide range of superstitions around the plant, for example about when it could be brought into the house and when it should be removed; by the mid-19th century, however, consumer culture was overtaking folk beliefs, and it was estimated that around a quarter of a million branches of holly were being sold annually for Christmas in London. In Europe, *I. aquifolium* had a surprising use historically as animal fodder, as it has some of the highest calorie and nutrient count of any forage. Spineless or young growth was favoured, and its survival and spread in pastureland was often encouraged. The wood could be used for a number of specialist purposes, while birdlime was made from a mucilaginous material obtained from the bark.

Herbal uses involving holly berries or bark are various. Among some Native American communities, *Ilex vomitoria* (yaupon) was used for ritual purgings; today it is sometimes consumed as a herb tea, as it is a stimulant. Many species in the Americas served as herbal medical treatments, particularly in treating fevers. The stimulant tea, maté, of South America is also a holly, *I. paraguariensis*.

Selection of distinct varieties for fruiting quality, fruit colour, and leaf colour and shape began in the 19th century. The gene pool of evergreen varieties is still dominated by *Ilex aquifolium*. The deciduous hollies are mainly grown for their spectacular and very persistent fruit; there has been much North American varietal selection and hybridisation between *I. decidua*, *I. verticillata*, and a number of others. Commercial production of both evergreen and deciduous hollies is as much for the floristry market as it is for the nursery trade.

In Japan, *Ilex serrata* has a long history in cultivation, but cultivars were usually given new names when introduced to the West, so their origin is obscure; this species has also been very popular for bonsai. *Ilex crenata* was also grown extensively in Japan, with many cultivars being produced during the Edo period; it was heavily promoted by the Arnold Arboretum in the 1930s and is now well established in cultivation, becoming especially popular from the 1990s onwards as a replacement for disease-hit boxwood.

Impatiens Balsaminaceae

A genus of around 1,000 species, almost entirely Old World, *Impatiens* is noted for a high level of local endemism (i.e., small isolated populations); there are 100 species in East Africa, 200 in Madagascar and Indian Ocean islands, 300 in the Himalayas and China. The Western Ghats of India and the highlands of Southeast Asia also have rich populations. The genus is very much one of the montane regions of warmer climates, where species tend to be clonal perennials; in cooler temperate zones—Europe and the Himalayas—species are more likely to be annual. Habitats are generally damp and shaded, including semi-aquatic and epiphytic locations. A few species, however, have a xerophytic character. Flower shapes are designed for particular pollinators, usually from among butterflies, moths, or birds; many species are also able to self-pollinate. The name actually does relate to impatience, as the seeds are distributed explosively. There is little natural hybridisation but much variation within species, which makes *Impatiens*, arguably, a genus with enormous horticultural potential, nearly all unrealised; *I. omeiana*, introduced in the 1990s, is a recent example of this potential, a reliably hardy perennial from China, grown chiefly for its foliage.

The Himalayan *Impatiens glandulifera* was introduced in 1838 to Europe as an ornamental annual, since when it has spread widely along river and canal banks to become one of Europe's most visible alien plants. By the end of the century it had naturalised itself on the Pfaueninsel, an island in the river Havel in Berlin; interestingly it no longer seems to be there, which offers hope that other areas in Europe swamped by the plant

A variety of *Impatiens balsamina* painted (in watercolour on vellum) by self-taught botanist and artist James Bolton (1735–1799). Images of balsams with multicoloured flowers are not uncommon from the 18th and 19th centuries; the plants are only rarely seen today. Courtesy of RHS Lindley Library.

may one day lose it. There is little evidence that it presents much of a threat to native species, but its suppression of common native perennials whose roots hold riverbanks together is a problem in some places.

The leaves of several species are eaten as a vegetable in Indonesia, the seeds of *Impatiens glandulifera* are pressed for oil in Nepal, while *I. balsamina* has some uses in Chinese traditional medicine and is used for washing hair in Vietnam.

Impatiens balsamina appeared in the first half of the 16th century, brought from India, although it is not clear by whom. The plant was little grown until a number of colour forms occurred in the 17th century; by the mid-19th it was extremely popular, although more so in continental Europe than in Britain. The rich and vibrant range of colours was particularly appreciated. The 20th century, however, saw a big decline in its use, almost to commercial extinction, as it was often displaced by other impatiens, notably varieties derived from the East African *I. walleriana*—actually a perennial in nature, whose shade tolerance relative to other annuals, long flowering season, and extremely compact growth habit have made it very popular since its introduction in 1883.

Impatiens hawkeri from Papua New Guinea was discovered in the late 19th century but not commercially developed until the 1970s, although apparently it was grown by native Papuans. The USDA and Longwood Gardens sponsored an expedition to the island in 1970, which was followed by a commercial breeding programme, with varieties being launched in 1989. With disease problems threatening the "traditional" gene pool, the use of new species in breeding is likely to increase.

Ipomoea Convolvulaceae

There are some 500 to 650 species of *Ipomoea*, whose name derives from the Greek for "worm," after the twining habit of many. The genus includes woody as well as herbaceous species and is widespread, generally in warmer climates. Species important in cultivation are annuals or herbaceous perennials from woodland edge and other transitory habitats. In some cases, particularly with the perennial climber *I. purpurea*, they have become notorious as invasive aliens.

Ipomoeas typically contain a high level of alkaloids, which discourages predation. Derivatives from some species, notably from South America, have found use in herbal medicine and as psychoactive drugs for use in religious ceremonies; a compound similar to LSD can be extracted from the seed of several species, including the popular morning glory (*Ipomoea nil*). A number of species, however, are benign enough to be eaten, notably the sweet potato (*I. batatas*) and the foliage of water spinach (*I. aquatica*).

Ipomoea purpurea and other South American species appeared first in the West, during the 18th century, with Japanese varieties being introduced after the opening up of Japan in the mid-19th century. *Ipomoea nil* was introduced from China to Japan in the Nara period; it became a very popular subject for artwork and a particular favourite of amateur breeders from all social classes. The plants are annual and can only be

A display of morning glories, forms of *Ipomoea nil*, at the Botanical Garden of Everyday Life, Sakura City, Japan, including one of the mutant forms (left). The museum preserves a wide variety of traditional cult plants and holds annual exhibitions for many of them.

propagated from seed, which appears to have stimulated considerable research into genetics; during the Edo period, procedures for selection became very advanced. Particularly prized were *demono* ("segregating") plants with bizarre leaf and flower shapes; these were sterile, however, and the genes' coding for these characters recessive. Breeders corresponded extensively, and there is clear evidence from publications that they had reached the same conclusions as did Mendel a hundred years later. As with several other species, one of the most rigorous schools of morning glory cultivation was in the region dominated by the Higo clan around the city of Kumamoto. Large-flowered varieties are still widely grown in Japan, with the mutant forms kept going by a few devoted collectors.

Iris <small>Iridaceae</small>

Iris is, after *Rosa*, the most genetically complex hardy plant genus. An overview of it presents several oddities:

- No other genus of cultivation includes plants from such totally different habitats, from desert to waterside marginals. There is considerable correlation, however, between taxonomic divisions and habitat preferences.
- Flower colour range is exceptionally broad—basically yellow or blue/purple in nature, but with pinks and many very dark purples to almost black now available (no true reds—yet!). Indeed, the genus is named after the Greek goddess of the rainbow.
- Very few natural species are in cultivation—nearly all garden plants are hybrids.
- Species or varieties with a broad habitat tolerance are few and far between; many are quite particular about conditions, or are relatively high maintenance. Most could be described as connoisseur plants.
- Flowering season tends to be short. No breeder has come up with a gene for long-flowering (again—yet).
- The flower shape is broadly universal, with standards (the true petals, standing upright in the centre) and falls (petal-like sepals facing down and out). The inner part of each fall is covered by an additional petal-like structure, the style arms, which have evolved from the style. There may (or may not) be a beard of hairs at the top of the fall.

The approximately 270 members of the genus are divided into six subgenera, which are then subdivided into sections. Subgenus *Iris* comprises six sections, five of which are the aril irises, an informal grouping of bearded species that are among the most extravagantly beautiful of all flowers. Members of section *Oncocyclus* are particularly exotic-looking—nearly all these arils are from the harsh continental climates of the Middle East and central Asia, and they will only really flourish in similar climates (we can all enjoy their genetic input into the bearded iris hybrids, however). The aril designation also covers sections *Hexapogon*, *Psammiris*, *Pseudoregelia*, and *Regelia*; most of these are central Asian semi-desert plants. The sixth section, *Iris* (or *Pogon*), includes less-demanding plants from dry environments in southern Europe and the Middle East; these are the ancestors of the approximately 10,000 cultivars of bearded irises, divided by enthusiasts into various categories defined by height.

Subgenus *Limniris*, the beardless irises, contains 16 sections. Of these, the most important for gardeners are *Californicae*, the Pacific Coast irises from the American Pacific Northwest; *Hexagona*, the Louisiana irises, moisture-lovers from the American

South; *Laevigatae*, which includes *Iris ensata* (the ancestor of the so-called Japanese irises) and several other moisture-loving temperate zone species; *Sibiricae*, which includes *I. sibirica* and several other familiar garden plants; and *Spuriae*, which includes a number of species that have been crossed to form a relatively easy garden group, the spurias.

Subgenus *Scorpiris*, the Junos, are like the arils very much desert plants from central Asia, similarly gorgeous and often difficult. Subgenus *Xiphium* (bulbous irises) includes several species with a long history in cultivation; subgenus *Hermodactyloides*, likewise bulbous, includes many dwarf species from the eastern Mediterranean eastwards, some of which are popular and easy rockery plants. Subgenus *Nepalensis* contains relatively obscure species.

All irises are herbaceous or semi-evergreen perennials, with a strong stress-tolerating tendency. These are plants that commit to the long-term: their rhizomes may not always be that persistent, but they are very good at steadily exploring and occupying space; *Iris sibirica*, for example, dies out in the middle of the clump but covers everything around it in a dense thatch of slow-to-decay leaf litter, suffocating competing plants.

The distinctive flower shape is easily stylised; the fleur-de-lis, based on the European *Iris pseudacorus*, was adopted as an emblem by a 5th-century Frankish king and is now a national symbol for France. Orris (the dried roots of *I. germanica* and *I. pallida*) has a long history in European and Middle Eastern herbalism and perfumery; currently it is used as a fixative in perfume and potpourri manufacture, as a flavouring in Arab (most notably Maghreb) cuisine, and as one of the ingredients of gin.

The front cover of the Seitaro Arai catalogue. The name used is the old one for *Iris ensata*. Courtesy of RHS Lindley Library.

Iris growing today is largely a connoisseur hobby, although in the appropriate climate, a group that requires much care and attention elsewhere may be treated as an "ordinary" low-maintenance garden plant. The few that do perform well in a wide range of gardens over many years without cosseting are either very early hybrids or selections, or species, e.g., *Iris unguicularis*, an old favourite for neglected spots against walls, or the waterside *I. pseudacorus*.

Bearded irises

Iris germanica was possibly introduced to western Europe by the Romans and was for certain in monastery gardens from the early Medieval period. *Iris pallida*, *I. variegata*, and *I. pumila* were all grown from Medieval times; drawings of herbal plants of the period record a wide variety of iris flower patterns. *Iris susiana*, an *Oncocyclus* species from Iran, was brought to Europe in the 16th century; known as "lady in mourning" in German, its dark colouration made a big impact.

Systematic breeding started in the 1800s in France and Germany. By 1830 there were 100 varieties in France, many bred by the chief royal gardener, H. A. Jacques. Slightly later, Jean-Nicolas Lémon (1817–1895) became a successful breeder—entirely through random natural crossing. His masterstroke: he named his irises after Greek and Roman gods, goddesses, and heroes; popular novelists and their characters; opera personalities—anything to capture people's attention. Messrs. Vilmorin started working with irises in 1865. Lemoine introduced over 100 cultivars; other breeders, including van Houtte, Salter, and later Perry, obtained his plants and then did their own work. There was something of an overlap with the daffodil trade, with the well-known English bulb dealers Thomas Ware and Peter Barr dealing in both. A key German company was Goos & Koenemann, many of whose plants were purchased by Americans of German origin, so giving U.S. breeders plenty to work with; many of their best varieties are still available.

English breeders began working with tetraploid species from the Middle East around 1900; these reached the United States around a decade later, stimulating an intense phase of breeding, and the foundation of the American Iris Society in 1920. Ornamental plant breeding in general was stalled in Europe during the 1920s, owing to postwar economic difficulties, and in Britain iris breeding in particular suffered an unexpected blow with the untimely death of W. R. Dykes (1877–1925), a grower and breeder who had written the classic study of the genus. American breeders, however, forged ahead, their pink varieties a particularly noteworthy breakthrough.

After World War II, more species were pulled into the gene pool, one of the aims being plants that would tolerate a wider range of climatic conditions. Particular progress was made with dwarf bearded iris; the genes for small size were brought in by *Iris pumila*, with Goos & Koenemann heavily involved. Botanists interested in iris combed southeastern Europe and the Middle East for new forms of familiar species to use in breeding. Among these have been the spectacular *Oncocyclus* species, which themselves have been the subject of breeding in Israel. Small-scale and amateur breeders have always been important, e.g., the English artist Sir Cedric Morris (1889–1982).

A coloured engraved plate of irises from a publication of 1776, *Collection precieuse et enluminée des fleurs les plus belles et les plus curieuses, qui se cultivent tant dans les jardins de la Chine, que dans ceux de l'Europe* by Pierre Joseph Buchoz (1731–1807), a French doctor and botanist. The species are not known. Courtesy of RHS Lindley Library.

Xiphium irises

These flourish only in Mediterranean climates. *Iris xiphium* (Spanish iris) was introduced to the Low Countries (present-day Netherlands, ruled by Spain at the time, and Flanders) in the 16th century, along with *I. latifolia* (English iris), distributed by English sailors (as *I. xiphoides*), hence the common name (although it too is Spanish). The Dutch resourcefully cultivated (and bred) these irises, alongside their tulips—all bulbs whose optimum growing conditions were remarkably unlike their cool and damp new home. The hybrids, or Dutch irises (as they came to be known), were turned into a year-round cut flower crop by van Tubergen in Haarlem in the early 20th century. Today these are the leading iris in the cut flower trade, grown in huge quantities under glass.

Spuria irises

These are derived from species growing on the southern edge of the temperate zone in Eurasia. *Iris graminea* has been known since the 1500s, but it was not until Sir Michael Foster (1836–1907), a prolific British breeder of all sorts of irises, turned his attention to them that much was done in developing this group. Breeding has continued, but interest in them remains modest.

Japanese irises

Iris ensata var. *spontanea* has traditionally been grown along the edges of rice paddies, its flowering a signal that the time has come to plant the rice. Selection started in the Edo period, with numbers building up over time to today's thousands of cultivars. The Higo clan had a particular interest in them; clan members had an iris society with strict rules—in particular, that members were forbidden to give the irises to outsiders, and when members died, the plants had to go back to the society. The Higo varieties were exhibited in such a way that the flowers were at the height of viewers seated on tatami mats. Interest declined during the later Edo, but U.S. interest encouraged commercial production during the Meiji era.

Iris laevigata, grown as dyestuff in China and Japan, was also cultivated in the Edo. Although occasionally it was featured in art (one of the most famous of all Japanese artworks is a screen by the Edo-era artist Korin Ogata, showing simply a huge clump of this species), relatively little breeding was done.

A colour woodblock print of *Iris ensata* 'Enbi-no-Waza' from the second volume of the catalogue of Seitaro Arai, a Yokohama nursery that specialised in this species. Dated c. 1890s. Courtesy of RHS Lindley Library.

Japanese irises thrive in the climate of the eastern United States, which is similar to that of much of Japan; innovative breeding has contributed to their increasing popularity.

Pacific Coast irises

These are small plants with distinct but not excessively difficult habitat requirements. They hybridise readily and produce flowers in a rich array of colours. Thriving as they do in one of the world's most culturally and economically dynamic regions, the Pacific Northwest, it is not surprising that they have become very popular with gardeners and amateur breeders. These irises too have crossed over into the art world. Oregon-based artist George Gessert breeds a number of genera but has a particular love of Pacific Coast irises; as part of his work, he presents collections of his hybrids at galleries as installations, inviting audiences to help him select plants (the implication being that those not chosen are destroyed).

Siberian irises

The very widely distributed *Iris sibirica* was known for centuries but rarely cultivated. Other species in the section *Sibiricae* arrived from Asia from the late 19th century onwards. Perry and others in Britain started to produce hybrids in the 1920s, including some crosses with Pacific Coast species. Modest efforts have continued since, with the main development being much fuller petals and almost horizontal falls.

Louisiana irises

Flourishing only in climates with warm summers and places with moist soils, the Louisiana irises are the newest group to be intensively hybridised. They have a distinctly voluptuous quality, with large, flat flowers in rich colours—the best oranges, pinks, and (almost) reds of the whole genus, as well as purples and lavenders. The centre of breeding has been the American South, with notable work accomplished in South Africa and Australia as well.

Hermodactyloides

Two species from this bulbous subgenus are particularly important in gardens: *Iris reticulata* and *I. histrioides*. The first came to Europe initially through the Berlin Botanic Garden in 1821, but it was not until the end of the century, when Dutch bulb growers started to make selections and crossing with the recently introduced *I. histrioides*, that the plants became fully commercial. Both are from the semi-desert or montane regions of the Caucasus and northern Iran but in cultivation have proved a rich source of easy-to-grow early-season irises in bewitching blues and purples.

Iris reticulata flowering in April with *Scilla ingridiae* var. *taurica* above Kemaliye, Turkey.

Jasminum Oleaceae

Jasminum (whose name derives from the Persian for the plant) contains some 200 species, a mix of true shrubs and true climbers (with twining stems)—and the rather awkward mid-category between the two. Most are plants of the woodland edge, native to tropical and subtropical Eurasia and Australasia, but several seep into cooler regions, including one native to Europe. The centre of diversity is southern and southeastern Asia. Many are pollinated by night-flying moths.

Jasmines are greatly esteemed in the Middle East for their fragrance and in the Far East as a flavouring for tea. Their essential oil has long been used in perfumery (Grasse in the south of France is a key growing area) and latterly in aromatherapy (as a de-stressor) and dermatology.

The white species known usually simply as jasmine (*Jasminum officinale*) has become a template for a great many other "jasmines," totally unrelated climbing plants with white, heavily scented flowers. Thought originally to have come from somewhere in a belt from Iran to northern India, the plant has been in cultivation so long that the date and location of its original introduction is lost in the mists of time. It possibly arrived in Europe at the time of the Medieval Crusades. Two other species were known from early times too, but being tender, only grown in frost-free climates: *J. grandiflorum* (widely confused with *J. officinale*) and *J. sambac*. *Jasminum sambac* is probably of Indian origin, as many varieties occur there; they are widely used for Hindu *darshan* (worship rituals), such as the making of garlands for statues of the gods.

The yellow winter-flowering *Jasminum nudiflorum* is one of the few other truly hardy species but lacks scent. Introduced by Fortune, who had found it in a nursery in Shanghai, it was not clear to gardeners what to do with it, as its habit is neither self-supporting nor truly climbing. For that matter, there had also been a long history of not knowing quite how to deal with *J. officinale*, with attempts at training it as a standard often proving fruitless. Jekyll suggested that it should be planted to trail over a large rock, to which Coats (1963) acidly adds: "for those who happen to have a large rock about the size of a small greenhouse conveniently to hand."

Jasminum polyanthum was brought over from Yunnan by Lawrence Johnston, creator of the famous Hidcote garden in the English Midlands, in 1931. Not sufficiently hardy to grow outside in northern Europe, it has become a fixture in the garden centre, mass-produced as a throw-away pot plant for the post-Christmas period.

"Common White Jasmine on entrance to covered avenue or pergola," from an 1887 copy of *Gardening Illustrated*; *Jasminum officinale* remains a favourite for rambling over garden structures.

Kniphofia Xanthorrhoeaceae

This genus of perennials, named after German botanist Johann Kniphof (1704–1763) of Erfurt, has also been known as *Tritoma*. The vast majority of the approximately 72 species are South African, with a few species making it further north, to montane and hilly regions in Ethiopia, Democratic Republic of Congo, and Rwanda; there are also two on Madagascar and one in Yemen. Most species are evergreen; those which are not become deciduous during the dry or cold season. Most are plants of damp soils, found along streamsides, in seasonal marshes, or in damp grassland, often in montane areas; many of these environments, however, are not necessarily wet all year round. The plants have had some medicinal use in South African cultures.

Plants, introduced to European botanic gardens from the early 18th century onwards, were treated as greenhouse subjects, until *Kniphofia uvaria* was planted out successfully at Kew in 1848. German botanist and entrepreneur Maximilian Leichtlin (1831–1910) raised several hybrids in the relatively warm summer climate of southwestern Germany. British gardeners discovered early on that they flourished in the mild damp climate of the western part of the country; in particular they took off on the west coast of Scotland in the 1860s, where they acquired the name of "Bailie Nicol Jarvie's poker," after a character in Walter Scott's *Rob Roy*. Robinson (1921) devotes several pages to them in *The English Flower Garden* but deems them not hardy enough for general use. Various hybrids were raised during this period, but little further breeding seems to have been done until the revival of interest in perennials in the 1960s. The last decade of the 20th century saw some more introductions of species from South Africa.

Kniphofia caulescens in a bog at 2,500m, near Rhodes, Eastern Cape, South Africa.

Lathyrus Fabaceae

Named after the Greek for "pea," *Lathyrus* comprises some 150 species of herbaceous plants scattered fairly evenly over Eurasia, East Africa, North America, and temperate South America. The main centre of diversity is the eastern Mediterranean region. Most are climbers, but some are ground-dwellers like the perennial *L. vernus*, a Eurasian woodland species. This latter is a true perennial, but many others are short-lived perennials or annuals, with a pioneer habit, typically growing in woodland edge or open conditions, sometimes as arable weeds. *Lathyrus tuberosus*, a European species, has edible tubers, apparently very nice to eat but only produced in small quantities. The "peas" of the genus often contain toxins and are not eaten.

The short-lived perennial *Lathyrus latifolius* was grown from the late Medieval period onwards, becoming very popular with country people in central Europe during the 18th century; however, it was to be almost totally displaced by the sweet pea: *L. odoratus* came originally from Sicily, where it had been grown since the late 16th century at least. It was introduced to Britain by Robert Uvedale (1642–1722), a scholar and teacher who corresponded with a Sicilian monk, Franciscus Cupani, who sent him seed—which came up in a variety of colours, indicating that it had indeed been in cultivation there for some time.

There was little sign of the flower's remarkable potential until the latter half of the 19th century, when deliberate hybridisation began to increase the number of varieties. Col. Trevor Clarke of Daventry in the English Midlands used the South American *Lathyrus magellanicus* to bring blue into the gene pool. In 1870 Henry Eckford (1823–1905), a head gardener for several houses in the area, began and then proceeded to dominate breeding, which he did as a retirement hobby. In 1900, 264 varieties were shown at the Bi-Centenary Sweet Pea Exhibition, held in the Crystal Palace. The National Sweet Pea Society was founded the next year, which also saw the first variety with a wavy edge to the petals, 'Lady Spencer', bred by Silas Cole, gardener to the Earl of Spencer. Another wavy-edged variety was found around the same time—'Gladys Unwin', by William Unwin, who

> A dinner-table decorated heavily with Sweet Peas spoils my dinner, as I taste Sweet Peas with every course, and they are horrible as a sauce for fish, whilst they ruin the bouquet of a good wine.
>
> (Bowles 1914)

had a business selling the flowers to London markets; the company he founded is still a leading producer and breeder. The flower went on to become very popular in Edwardian Britain, particularly as a table decoration. By this time the sweet pea had also become very popular in several other European countries and in the United States, Eckford having supplied seed to clients in Boston.

Gardeners and nurseries on the East Coast started breeding sweet peas almost as soon as they arrived. For much of the 20th century, a leading breeder was Anton Zvolanek of New Jersey; a particular strength of his company was the production of multistem and semi-dwarf varieties. California breeders, including those employed by Burpee, began working with the plant in the 1910s; the state's Central Valley once grew fields of the plants for seed production, but the shift to customers buying starter packs of young plants rather than seed has greatly reduced the acreage. Modern breeding has focussed on disease-resistance and dwarf tendril-less varieties for small-scale use or for containers. As a whole, however, the plants have undoubtedly declined in commercial importance since their early-20th-century heyday. They are still grown for competitive showing by amateurs but on nothing like the scale of former years.

Lavandula Lamiaceae

Lavandula (from the Latin "to wash") is found from the Atlantic Isles through the Mediterranean region, North Africa, and the Middle East, with a small disjunct population in India. All the 39 species are subshrubs. With a lifespan which (in cultivation) rarely exceeds 20 years, lavenders occupy an ecological role similar to that of many subshrubs from semi-arid Mediterranean climates: they are part of a pioneer community that takes over after deforestation or fire. Lavender's oils speed the decomposition of leaf litter, and plants themselves have been shown to act as a nurse plant for cedar regeneration. They are of major importance for pollinators.

Long esteemed for its scent, lavender was historically grown as a crop plant, its foliage destined to be dried or tipped into the still to produce oil. It was widespread in the 16th century in France and the German-speaking lands, in monastery and apothecary gardens. Lavender is incredibly versatile as a medicinal herb, being made into powder, water, oil, syrup, sugar, paste, and a spirit. Its scent has been widely used to deter moths among stored clothes and textiles, as well as in herbal veterinary medicine. It is still used for its sedative, anti-inflammatory, and antiseptic effects, as well as for pain relief, treating colds and fevers, skin problems, delousing, and other insecticidal functions. Currently it is enjoying a revival in food, in both savoury and sweet dishes, and has very extensive usage in personal care products.

The Greeks and Romans primarily grew the more tender *Lavandula stoechas*. Hildegard von Bingen was the first to mention lavender in Europe, and during the Medieval Warm Period it was grown extensively in monastery gardens. In the

A coloured engraving of *Lavandula stoechas* from the sixth volume of *Flora Graeca* by John Sibthorp (1758–1796) and drawn by Ferdinand Bauer. Sibthorp was a professor of botany at Oxford and travelled extensively in southeastern Europe from 1786 until his death. Courtesy of RHS Lindley Library.

"The Lavender Garden," a watercolour by Edith Helena Adie (1865–1947). Adie was a well-known painter, specialising in landscapes and garden views. Around 1922–23 she painted a series (of which this is one) of the gardens at Dyffryn House, South Wales, commissioned by the owner, Reginald Cory, a passionate plantsman and great benefactor to horticulture. Courtesy of RHS Lindley Library.

post-Medieval period, there were multiple introductions to northern Europe. *Lavandula stoechas* has been in cultivation in Britain, off and on, since the 14th century, and continuously in France. Fields of it were grown in early-19th-century England, for medicinal, culinary, and cosmetic purposes, and southern France remains a major area for commercial production. Its introduction to North America was initially through California, by Spanish Franciscan monks.

Lavandula angustifolia (English lavender) is the "default" lavender of northern Europe; its common name is presumably a recognition of a long history in that country. The horticulturally important *L.* ×*intermedia* arose naturally between *L. angustifolia* and *L. latifolia*; it produces larger quantities of oil, with a higher proportion of camphorous compounds. The cross has been made several times over the last few centuries. *Lavandula dentata* (French lavender) was first described by Clusius, although it had long been cultivated by the Arabs; it is particularly widely grown in Australia, as it flourishes in hotter climates.

Although occasionally used for edging from the 18th century onwards, the use of lavender in gardens does not really take off until the 20th century, when it was seen as a distinctly Old World plant. Gertrude Jekyll's promotion of 'Munstead' as a dwarf hedging plant, in the 1900s, was symptomatic of its growing popularity. Many varieties originated in southern Europe (e.g., 'Hidcote', brought from the south of France by Lawrence Johnston in the 1920s).

Leucanthemum Asteraceae

Forty Eurasian herbaceous perennials make up this genus, whose name derives from the Greek for "white flower." It has, over the years, been redefined by botanists several times. All are clonal perennials, of a distinctly competitive or pioneer-competitive character. They are mostly plants of grassland. *Leucanthemum vulgare* (ox-eye daisy) is a well-known component of wildflower seed mixes in Europe, exuberant in the first few years before (usually) declining to insignificance. Traditions in several parts of Europe link *L. vulgare* and other species to mid-summer celebrations but also grant it the power to keep away thunderstorms, so bunches of it were sometimes hung on farm buildings.

Leucanthemum ×*superbum* (Shasta daisy) is among the most familiar perennial garden flowers, and one which shows a great ability to survive neglect. It is the crowning achievement (for the domestic gardener) of that most flamboyant of plant breeders, Luther Burbank, who claimed to have made this particular cross from two European and one Japanese species; later work, however, has suggested that only two species were used, or possibly just the one, *L. lacustre* (he famously did not keep records). Burbank raised more than 100,000 seedlings over a six-year period, constantly selecting those which had the biggest flowers and those which bloomed the longest; he named the resulting plant after a peak visible from his California farm, Mount Shasta. Over the years, the plant has thrown up a number of mutations, which have provided some useful variation. There is one good story, from England, about nurseryman Horace Read, who was on a train journey sometime in the 1920s, spotting a semi-double form of the plant growing on an embankment. On his return journey he pulled the cord for the emergency brake, leapt off the train, and dug up the plant. He used the plant in further breeding, giving the Shasta daisy a second lease of commercial life.

"If you like flowers that are huge . . ." begins the sales blurb in the 1946 Kellogg's catalogue, for a variety with 12cm blooms. Maybe the public did not, as this cultivar of *Leucanthemum* ×*superbum*, the Shasta daisy, seems to have disappeared. Many others have survived, however, and the plant has proved a true perennial favourite.

Lilium Liliaceae

The name is from *leírion*, the Greek for *Lilium candidum*. The genus comprises around 150 species—all herbaceous bulbous perennials, although *L. candidum* and a few others produce an overwintering rosette of ground-level leaves. All are northern hemisphere, with centres of diversity in eastern Asia and western North America. Overwhelmingly these are plants of cooler, often montane, regions, from woodland edge or higher-altitude grassland habitats, moist cliffs, and rocky situations—as is evidenced by *L. auratum* being traditionally grown on cottage roofs in Japan. Some species go as far south as the subtropical zone, and *L. candidum* is of eastern Mediterranean origin, but its rarity in the wild indicates perhaps that its habitat has been largely lost.

Botanists currently recognise eight sections of the genus; however, since relatively few species are currently cultivated, the seven divisions into which the cultivated species and hybrids are grouped are of more relevance for gardeners:

- Asiatic hybrids (Division 1). Derived from central and east Asian species, with medium-sized, upright or outward-facing flowers, not usually fragrant.
- Martagon hybrids (Division 2). Derived from *Lilium martagon* and others with a Turk's-cap shape.
- Candidum (Euro-Caucasian) hybrids (Division 3). Derived largely from European species.
- American hybrids (Division 4). Derived from North American species, often clump-forming and tall.
- Longiflorum hybrids (Division 5). Derived from *Lilium longiflorum*.
- Trumpet lilies (Division 6). Derived largely from *Lilium regale* and other Asian species, with trumpet-shaped flowers, often strongly fragrant.
- Oriental hybrids (Division 7). Derived from *Lilium auratum* and related east Asian species, often large plants, with outward-facing, star-shaped, strongly fragrant flowers.

In horticulture, divisions 1, 6, and 7 are commercially the most important, and 5 too, but primarily for floristry. Just mentionably, the word "lily" has been used for a wide

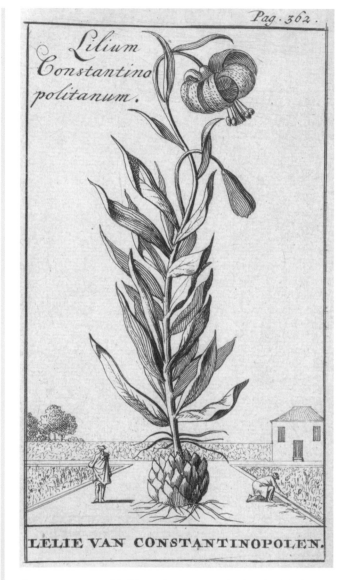

A martagon-type lily "from Constantinople" by Van Eeghen (Blankaart 1698); most probably *Lilium chalcedonicum*, which arrived in Europe from the Ottoman Empire during the 16th century. Courtesy of the Rijksmuseum, Amsterdam.

range of unrelated monocot plants with vaguely trumpet-shaped flowers (e.g., lily-of-the-valley, day-lily, waterlily, arum lily), causing much confusion for florists and the non-botanist public alike.

Lilies would appear to be potentially long-lived plants. But—there are many "howevers." Lilies are distinct for having unprotected bulbs composed of scales, very vulnerable to drying out and damage. Most have roots emerging from the bulb base and from the stem. For a plant to survive from year to year, a daughter bulb has to grow within the old bulb to regenerate it for the next year; this may happen within the old bulb, but in American species, the new bulb is formed outside the old one, so the plant gradually moves; some American species are also stoloniferous, e.g., *Lilium pardalinum*. The level of increase from year to year varies between species but never appears to be that high. The plants do appear to be very sensitive to environmental conditions and so may not regenerate every year in cultivation. Quite possibly, they may fail to do so in natural conditions frequently, too, as there is a tendency to seed profusely. The fact that some species (e.g., *L. regale*) are able to flower from seed in as little as two years does indicate some pioneer character.

The upshot of all this is that lilies are nothing like as robust in cultivation as many bulbs; they have always had a reputation for being difficult, fussy, or (probably incorrectly) short-lived. They also suffer greatly from an aphid-borne virus, which has done much to decimate our heritage of them. The advantage of many hybrids is that they regenerate well from year to year, possibly a result of hybrid vigour.

Lilium candidum (madonna lily) is one of the oldest plants in cultivation. Grown and distributed by the Romans, and later by Christian cultures, the purity of its whiteness gave it great symbolic power: it was an emblem of John the Baptist, St. Joseph, St. Dominic, and St. Anthony of Padua, but above all, of the Virgin Mary (hence the common name). Widely known during Medieval times and grown during the early modern period, it is very frequently depicted in religious art and often appears in popular religious stories. On a more earthly plane, lilies were commonly grown as a starchy vegetable. In the mid-20th century, *L. pardalinum* was grown by California Chinese for consumption, and several species, including *L. auratum*, are still grown commercially as food crops in eastern Asia. Lily bulb scales are often eaten in soups, and the author fondly recalls lily bulb raclette at a food stall in Hokkaido.

Lilium auratum, *L. speciosum*, and *L. rubellum* were introduced to Europe by von Siebold around 1830. British growers started breeding using the Japanese gene pool soon after introduction; Dutch commercial growers joined in somewhat later, and it is thought they added the European *L. bulbiferum*. Now almost unknown as a garden plant, *L. bulbiferum* was one of the most common cottage garden plants in 19th-century

Front cover of the 1916–17 "Descriptive Catalogue of the Yokohama Nursery Co, Ltd." The company, which probably acted as a marketing organisation for many small nurseries around Tokyo Bay, had, by this time, offices in Japan, London, New York, and Shanghai. *Lilium auratum*, *L. speciosum*, and *L. longiflorum* are shown. Courtesy of RHS Lindley Library.

The Japanese imagery for selling lilies was still important in the 1950s. From the 1959 Stassen catalogue, clockwise from top: *Lilium* 'Fire King', *L. speciosum* 'Rubrum', *L. japonicum*, *L. henryi*, *L. auratum*, *L. regale*. It is interesting that few hybrids are offered and that *L. japonicum* is now very hard to get hold of.

Germany, along with *L. martagon*; they have been displaced in cultivation by Asian species, or more accurately, hybrids with a largely Asian component in the gene pool. A pioneer hybridist was Henry Groom in London in the 1840s; Charles Mason Hovey in Boston bought bulbs from him and pioneered hybridisation in the United States from 1846 onwards. Populations of wild *L. auratum* astonished early travellers in Japan; and during the Meiji era, vast quantities, all nursery-grown, were exported to Europe and North America, along with other species, notably *L. lancifolium*. This production was probably a shift in selling an already widely grown vegetable crop to clients who offered more money to admire them rather than eat them.

> Next to the Rose, there is not a fairer floure than the Lilly, nor of greater estimation.
>
> (Pliny the Elder, in Philemon Holland's 1601 translation, quoted in Woodcock and Stearn 1950)

Lilium regale was introduced in 1904 by Wilson, from the one valley where it grew in the Tibet/China border region; its arrival was described by one contemporary author as "epoch making," and in 1910, when Wilson returned to collect more, he suffered severe injuries in a rock fall. Now the most widely grown pure species, it has almost entirely displaced *L. candidum*.

There was considerable interest in lilies in the 19th century but also much frustration in growing them; Henry John Elwes's *Monograph of the Genus Lilium* (1880) brought together the accumulated experience of many growers and marks the beginning of a more confident phase. Hybridisation seemed to make the lily more tractable as a commercial and garden plant. There is little natural hybridisation, but it is not difficult to achieve in cultivation. The Japanese had undertaken some crossing during the Edo period using *Lilium dauricum*, producing flowers ranging from red to yellow, with spotted and unspotted variants; many were dwarf and ideal for growing in pots. They also worked with *L. auratum*, *L. speciosum*, and *L. rubellum*, essentially establishing the groundwork for divisions 1, 6, and 7.

A number of groups of hybrids (some of them seed strains rather than groups of true cultivars) have been produced over the years, but many are now lost, or found only in specialist collections. The Backhouse Hybrids were raised by Robert and Sarah Backhouse in the early 20th century from *Lilium martagon* and *L. hansonii*; their house in Herefordshire still has self-seeding lilies in the garden. The Bellingham Hybrids strain was developed from several American species at the USDA Bulb Station in Bellingham, Washington, during the 1920s; these almost certainly built on the work of Burbank, whose adventurous breeding efforts involved planting fields of lilies so extensive that their fragrance could be detected many miles away. Although widely grown at the time, his hybrids are now all lost to virus. This coupled with Burbank's slightly dubious reputation—he is said to have crossed a lily with a trillium—makes it difficult to assess his achievements.

Isabella Preston (1881–1965), Canada's first female hybridist, worked with *Lilium regale* at the Ottawa Central Experimental Farm—many of today's trumpets are descendants. She introduced the Stenographers series, named for typists at the institution, and the Fighter Group, named after World War II Allied fighter aircraft. The most successful breeder was Jan de Graaff (1903–1989), who began his first lily experiments in 1938 at the Oregon Bulb Farms, having come from Leiden, in the Netherlands, from a family that had been in the wholesale bulb business since 1723. In 1941 he produced 'Enchantment', a dark red with upright-facing flowers, the most commercially successful lily of all time.

Contemporary breeding is dominated by the needs of the cut flower trade, with an emphasis on characteristics that are not necessarily relevant to gardeners, such as upward-facing (and therefore rain-gathering) flowers. Biotechnology has enabled many difficult crosses to be made, so the gene pool is increasingly wide; it has also enabled some virus-ridden heirlooms to be "cleaned up" and commercially regenerated.

'Mrs. R. O. Backhouse', an early-20th-century hybrid, one of several between *Lilium martagon* and *L. hansonii*, known as the Backhouse Hybrids. This plant received an Award of Garden Merit from the RHS in 1921 and is the only one of the group that remains commercially available; it is still growing at Sutton Court, the Backhouse family home in Herefordshire, England, where they were produced.

Lilyturfs Asparagaceae

Liriope (named for the mother of Narcissus in Classical legend) includes six perennials, which unusually are true evergreens, i.e., the foliage lives for more than one year. The very similar *Ophiopogon* ("snake's head" in Greek) includes 67 species. The shared common name of the genera, lilyturf, evokes their potential as a grass alternative (and the fact that they used to be in the Liliaceae). All lilyturfs are native to eastern Asia. They are stress-tolerant plants that are well-adapted to life on the forest floor of woodland there, as being evergreen they can photosynthesise during the (usually mild) winter. Growth in hot humid summers can be rapid but is less so in regions with cool summers. The plants have an ability to survive for years when conditions are poor, marking time until higher light or moisture levels on the forest floor enables them to put on growth which can be in the form of tight clumps or a more extensive guerrilla spread (at least in some ophiopogons). Flowering tends to be at the end of the summer, as temperatures cool, a common feature of plants from this region. There are minor medical uses in several east Asian herbal traditions.

Information on how these relatively humble plants first arrived in cultivation is scant. They have almost certainly been used in Chinese, Korean, and Japanese gardens for centuries, with varieties being selected on the basis of foliage colour and habit, and appear to have been introduced to the West during the 19th century. They have been hugely successful in those U.S. regions where summer climate is similar to that of their Asian home; in the South they have become very important as ground covers and edging.

Lobelia Lobeliaceae

Named for Flemish botanist and apothecary Matthias de l'Obel (1538–1616), *Lobelia* includes around 300 species, found mainly in the warmer parts of the world, with only a few temperate zone species. There are both woody and herbaceous species, long-lived and annual—the genus is amazingly disparate, with similar flower structure often the only defining characteristic. The origin of *Lobelia* would appear to be African, the plants' wide geographical dispersal over geological time helped by the very small seed. Hawaii is a remarkable centre of diversity, with 125 species of *Lobelia* and closely allied plants, all thought to have evolved from a single ancestor within the last 13 million years. The most extraordinary lobelias can be found in African and South American mountain regions—giant unbranched shrubs with leaves arranged in rosettes. It is difficult to summarise the preferences of such a diverse genus, but on the whole, plants seem to prefer open habitats with higher levels of soil moisture. The hardy species in cultivation have a tendency to be only weakly clonal and thus rather short-lived.

Lobelia inflata, which is not grown as an ornamental, was widely used for its psychoactive properties by Native North Americans. *Lobelia siphilitica* and *L. cardinalis* have been used to treat syphilis.

When introduced in the early 17th century from North America to France, *Lobelia cardinalis* (cardinal flower) caused a sensation, the scarlet colour of its flowers likened to that of a Catholic prelate's clothing (hence the common name). Nineteenth-century growers began to produce hybrids between this and several other species, including *L. siphilitica*, but nothing of major importance ever resulted. Odd attempts at breeding have continued ever since; arguably nothing is as good a colour or a garden plant as straight *L. cardinalis*.

Far better known are the small bedding lobelias derived from *Lobelia erinus*, found wild in the Cape and first grown in Europe at Leiden Botanic Garden in 1687. It was developed as a bedding plant in the second half of the 19th century, as it proved ideal for the contemporary fashion of carpet bedding; unlike many of the plants used for this style, it has continued to be of major importance. In German it is known as *Männertreu*—a humorous comparison between the short-lived flowers and the supposed loyalty of men.

Lonicera Caprifoliaceae

The name honors Adam Lonicer, a 16th-century German botanist. There are 120 species in the genus, of which nearly 100 are found in China, the remainder scattered over the rest of Eurasia and North America. Species fall into two distinct subgenera: *Chamaecerasus* and *Lonicera*—the former mostly shrubs, the latter climbers, with weakly woody growth. All seem to be very much plants of the forest margins. Although a few of the warmer-climate species may get large enough to scramble up full-grown trees, most take advantage of woodland glades or woodland edge habitats to get to the light with less effort. The shrubby species are also plants of woodland edges but also often form a major part of the scrub communities of slopes too steep for trees, environments verging on the semi-arid or where grazing limits tree growth. The evergreen shrubby species are generally plants of the forest floor. The climbing honeysuckles generally attract moths as pollinators, hence the evening and nighttime scent, or, in North America, hummingbirds—if the latter, the flowers are more likely to be scentless and orange or red in colour.

The twining nature of climbing honeysuckles has probably driven their popularity as love tokens in Europe. There have been minor medicinal uses, mostly in the Far East. The stems have been used for basketmaking by Native Americans and in the Far East. *Lonicera caerulea* fruit are good to eat, although usually cooked as jam and filling for tarts. It was one of the crops promoted under communism to diversify the range of fruit available in Russia; in the post-Soviet era, selected varieties have become available in the West.

Popular with the Elizabethans but not always as a climber, *Lonicera periclymenum* would sometimes be trained up a pole and then kept as a standard. The origins of the old forms 'Belgica' ('Early Dutch') and 'Serotina' ('Late Dutch') are obscure and must date back to at least an importation to England by Flemish growers in the early 18th century.

American honeysuckles, mostly bright but scentless, were first introduced by the younger Tradescant. Several Asian species arrived during the 19th century, including *Lonicera tragophylla*, which was crossed in the garden of the Royal Hungarian Horticultural School in Budapest with American *L. sempervirens* to produce *L. ×tellmanniana*, distributed by the German Späth company in 1927. Another Asian species, *L. japonica*, has become a much-hated invasive alien in parts of the United States, as it smothers native vegetation.

Some shrubby species are valued for their sweetly scented flowers. All are from Eurasia, the best known probably the winter-blooming *Lonicera ×purpusii*, a hybrid between two Chinese species introduced by Fortune in 1845. Others are valued instead for their functional qualities, such as another Chinese evergreen species, *L. nitida*, introduced by Wilson in 1908, which rapidly became a popular hedging plant in the interwar years in Britain, quick-growing and dense enough to make into topiary but generally held to be well inferior to box, socially as well as horticulturally. The somewhat more sprawling Chinese *L. pileata* has become very popular as a "green cement" plant with the landscaping industry, and consequently much reviled. Somewhat rarer these days is *L. tatarica*, which was a very common garden plant in early-19th-century England.

Lupinus Fabaceae

The name comes from the Latin for "wolf," as Classical authors believed the plants depleted soil fertility through their vigorous growth; in fact the opposite is true, as these, like nearly all pea family plants, add nitrogen to the soil through the bacteria in nodules on the roots. Around 270 species, including annuals, perennials, subshrubs, and small trees, are found across Eurasia, North Africa, and the Americas. DNA evidence suggests the genus underwent two distinct periods of rapid evolution in the Americas—indeed, one of the most remarkably fast and extensive periods of evolution of any plant group so far studied. The presence of ancestral Old World lupins in the Americas dates back to around 15 million years, species unique to North America 11 million, to South America 2.5 million, with some species taking only 1 million to evolve.

Two regions experienced bursts of "evolutionary radiation," which is when new opportunities are presented by geographical or ecological circumstances, and a particular species reacts by rapidly evolving new species to make the most of the possibilities. The classic example is Darwin's finches on the Galapagos. The mountains of western North America and of the Andes were the opportunities offered to ancestral lupins. Species evolved to fill a wide range of habitats from Alaska right down to the tip of Chile, particularly high altitude and dry ones, including sand dunes, scree slopes, and the savage "cold desert" of the high Andean puna. To achieve this, lupins evolved into a remarkable range of forms, from ground-hugging subshrubs and small trees to rosette-forming perennials and annuals, and everything in between.

Several European annual species were grown by Romans as cattle fodder and for green manure. Lupin seeds have been put to various uses: they have been eaten like beans (although they often contain bitter chemicals that must be removed by thorough soaking); they have been used as a coffee substitute; and an oil expressed from them has been used to make soaps and ointments, including what amounted to an 18th-century face cream.

The species and hybrids of lupin in cultivation are all short-lived non-clonal perennials. They are easy to propagate from cuttings, but gardeners usually grow them from seed. Their being short-lived reflects the pioneer habits of their ancestral wild species; generally they are found in the earlier stages of succession, or in gaps that appear in existing grassland or scrub. *Lupinus arboreus* is found typically on coastal sand dunes, *L. polyphyllus* in wetlands or on riverbanks.

Perennial lupins from eastern North America began to appear in the 17th century and some of the many western species a century later, including *Lupinus arboreus* and the blue *L. polyphyllus*, found by David Douglas in 1826. It is thought that these two were the original gene pool for George Russell (1857–1951), who effectively launched the genus as a garden plant in the early 20th century. Russell was a northern English professional gardener who, having been given some lupins by an employer, felt inspired to grow better ones; he began breeding them on two rented allotments and staged his first exhibit at the Chelsea Flower Show in 1938, aged 79. Unlike many plant breeders, he was never afraid of the limelight, and he died at the age of 94, having had a very active retirement. Lupins rapidly became fixtures of the English herbaceous garden style and of working-class gardens but fell out of fashion from the 1960s onwards. New breeding picked up again in the 1990s.

Lupinus sp. in *Pinus ponderosa* forest, a typically very open woodland community, in the western foothills of the Northern Rocky Mountains, Washington State.

Magnolia Magnoliaceae

Named for French botanist Pierre Magnol (1638–1715), *Magnolia* is a famously old genus. It is also hard to define; current thinking suggests 200-plus species. Some botanists favour splitting it, others dumping the entire Magnoliaceae into it. The fossil record is rich, with family members dating back 98 million years, and *Magnolia* itself being of Cretaceous origin, i.e., contemporary with the dinosaurs but before bees: it is thought the first magnolias were pollinated by beetles. As might be expected from such an ancient plant, distribution is very disjunct, scattered through South America from southern Brazil up into eastern North America, and Southeast Asia up into China and Japan, with some presence in southern India. Southern China and Southeast Asia are a centre of diversity for hardy species, South America for tropical.

Magnolia is divided into three subgenera, 12 sections, and 13 subsections. The majority of horticultural species are in section *Yulania*. There is also now near consensus that the largely tropical genus *Michelia* should be included in *Magnolia*. All magnolias are trees, mostly large, of mature forest; they are slow-growing and long-lived and always a minority component. They tend to favour higher-rainfall and warmer-climate regions, although few are true rainforest species; many are found in montane regions.

Magnolia flower buds are used for treating sinus problems in Chinese herbalism, and the bark (primarily of *Magnolia officinalis*) is prescribed for a wide range of conditions, although it does contains high levels of alkaloids and concerns have been raised over its safety. Flowers are used for flavouring, or at least ornamenting, certain Chinese teas, while buds are deep fried and eaten as a spring delicacy; they are somewhat tasteless, but the petals make an attractive addition to a salad. Wreaths of *M. grandiflora* foliage are traditional in the American South (they dry slowly and can be kept for months), and the flower, an emblem of the Confederate army in the U.S. Civil War, remains a potent symbol of the white American South, occasionally being used for a more aggressive political stance (e.g., the *Free Magnolia*, the quarterly tabloid of the unrepentant League of the South). The 1989 film *Steel Magnolias* conjures up an image of something beautiful but also

> [T]he most delightful [scent] that can be conceived; far exceeding that of the rose; in strength equal to that of the jonquil or the tuberose, and far more delightful.
>
> (Cobbett 1829)

very tough; the tree symbolised the character of the women in the film, and by extension, the character of the women of the white South.

The first of many magnolias to make an impact in Europe was the North American *Magnolia virginiana*, sent over to London in 1688. The much more spectacular *M. grandiflora* was first grown in British gardens in the 1730s but a hard winter in 1739 killed most of them. Perhaps it was this experience that led a later generation of gardeners to treat them as wall shrubs; many of the most venerable today clad distinguished late-18th-century houses. The plant varies considerably in the shape and size of its foliage, and the time it takes to get round to flowering—early-flowering selections were being made by mid-19th century.

Magnolia grandiflora selections, and indeed all selections of the larger magnolias, have tended to be chosen for flowering at a young age. This species is an interesting one: it is a hexaploid (i.e., with three times the normal chromosome count), indicating that the plant is a legacy of natural hybridisation,

A hand-coloured engraving of *Magnolia grandiflora* (as *M. altissima*) drawn by Mark Catesby (1682/83–1749) and published in his *Hortus Europae Americanus* of 1767. Courtesy of RHS Lindley Library.

itself possibly the reason for the species' survival as successive ice ages would have driven many species to extinction. The plant has proved remarkably successful in horticulture, possibly even the most successful North American tree, being grown globally in both warmer and cooler climates. Hybridisation with the smaller *M. virginiana* and variety selection have resulted in a range of cultivars with different foliage characteristics.

Magnolia denudata (yulan) was the first Asian species to make its way westward. The Chinese had cultivated it since the Tang dynasty, often as a bonsai. Early experiences in Britain led many to consider it less than hardy, while its habit of flowering before the leaves open was seen by some as rather unattractive.

The most common magnolia in temperate zone cultivation is *Magnolia ×soulangeana*. An accidental hybrid it may have been, but its appearing in the early 1820s at the château of Étienne Soulange-Bodin, a French military man who devoted himself to gardening upon retiring, guaranteed its rosy future, and reliable flowering and a compact habit have ensured that the cross (*M. denudata* × *M. liliiflora*) has been remade many times. There are now many distinct cultivars. More compact still and immensely popular is *M. stellata*, introduced from Japan by Veitch's in 1878. The tree-sized early-flowering beauties, such as *M. campbellii* and *M. sprengeri*, arrived during the heyday of the plant hunters in the late 19th century. They remain the most aristocratic of garden plants, needing large gardens and climates with the minimum of untimely frosts.

Recent magnolia breeding has focussed, not surprisingly, on smaller-growing varieties for garden use, and on producing a truly yellow deciduous one. The latter objective has largely used North American *Magnolia acuminata* to provide the gene for flower colour.

Matthiola Brassicaceae

Named for Pietro Mattioli (Matthiolus), a 16th-century doctor and naturalist from Siena, Italy. There are around 50 species in this genus, which like many in the family, is made up of short-lived perennials that verge on the subshrubby. They are found around the Mediterranean, into North Africa, and across to central Asia. Most would appear to be short-lived pioneer plants of seasonally dry open habitats.

Stock is one of those common names that covers a multitude of confusion: it is used for several related members of the Brassicaceae (e.g., *Malcolmia maritima*, Virginia stock, which despite the common name is from southeastern Europe), all notable for their sweet smell, which reminded our ancestors of the gillyflower or pink; but since they had a definite stem, or stock, they were dubbed stock gillyflowers. The flowers are edible and make a decorative finish for a salad. There are some very minor uses in herbal medicine.

Matthiola incana is regarded as the true stock. Being both colourful and heavily scented, much early gardening effort was put into growing it, and by the early 18th century it was available in varied colours. 'Brompton' and 'Ten Week' were names originally applied to strains grown at a nursery at Brompton in London in the 1750s; they are now rather loosely applied to various seed strains. In the 19th century, the plants became very popular with the weavers of Upper Saxony, who, in order to minimise cross-pollination, would have each village organised to produce seed of only one colour of flower. Doubles, always the most sought-after form, have been around since the 16th century; but they are sterile and have to be grown from seed from parents that both carry the recessive gene for doubling. Danish seedsmen and the geneticist Øjvind Winge (1886–1964), best known for his work with yeast, improved the level of production of doubles considerably, making them a more worthwhile commercial proposition.

Matthiola longipetala (night-scented stock) is grown for its evening scent. None of the stocks flowers for very long, which has limited their commercialisation, and they remain largely a plant for small-scale amateur growing.

Meconopsis Papaveraceae

Until recently this was a genus of 43 species, but some now accept *Meconopsis cambrica* (Welsh poppy) of Europe's Atlantic fringes as *Papaver cambricum* (indeed, the name *Meconopsis* derives from the Greek for "poppylike"). *Meconopsis* species are all to be found in the foothills of the Himalayas, mostly on the wetter, eastern, Chinese side. *Meconopsis cambrica* is possibly a Cenozoic relic from when *Meconopsis* or an ancestral group once covered the whole of Eurasia, aridification then causing extinction and separation of everything between the Cambrian and Himalayan mountains. With the exception of *M. cambrica*, which can be almost weedy in some gardens, *Meconopsis* is one of those genera whose members are seen very much as connoisseur plants, romantic species with a good deal of legend attached.

Meconopsis includes clonal perennials, shorter-lived non-clonal species, and several which are monocarpic. Some species show considerable variation in longevity, which adds to the frustrations of growing them. They are plants of alpine meadows, open woodland, and woodland edge habitats in a monsoon climate, often at considerable altitude, where winters are cold enough to cause a complete cessation of growth. Their reproduction is characteristic of pioneer plants: they constantly regenerate themselves in short-lived gaps in more persistent vegetation.

The era of the great plant hunters saw the start of the meconopsis myth. That blue poppies existed created a sensation, particularly when *Meconopsis betonicifolia* first flowered in cultivation in 1924, although it had been discovered by Delavay in Yunnan back in 1886. Most species arrived in cultivation in the late 1800s but have never become widespread. Plants need cultivation conditions that approximate those of their natural habitat; therefore, really successful growers are rarely to be found beyond Scotland, Scandinavia, or the northern end of the Pacific Northwest. The spectacularly coloured flowers—pinks, reds, and yellows as well as the blues—and the enormous hairy rosettes of the monocarpic species have a strong appeal, but their fussiness in cultivation has resulted in them being plants that are mostly enjoyed in other people's gardens.

Monarda Lamiaceae

Monarda (named after 16th-century Spanish botanist Nicolas Bautista Monardes) comprises 16 species of herbaceous perennials from North America. Monardas are generally plants of open sites: waste ground, prairie, and woodland edges, with some species favouring damper ground, others drier; they are most common in the earlier stages of succession. Some species are very short-lived, but those in cultivation are potentially long-lived perennials. The gene pool carries considerable variation in persistence. Some species form short-lived shoots, which in the wild are able to exploit gaps in vegetation, but in garden conditions, this can mean that they are not able to regenerate and so die out. Modern selection, naturally, has focussed on plants with the genetic predisposition to form more persistent clumps.

> [E]verything about the plants is attractive.
>
> (Gerritsen and Oudolf 2000)

The nectar-rich flowers attract a very wide range of pollinators (hence the common name of bee balm). The aromatic foliage has been much used in herbal medicine by both Native Americans and settlers, for a range of minor complaints. Early settlers made tea from the plant, leading to another common name, Oswego tea. The leaves have also been used for flavouring black tea; yet another vernacular name, bergamot, alludes to their aroma, which is similar to the bergamot orange that flavours Earl Grey tea.

First described from France's North American colonies in the 17th century, species of *Monarda*, at first thought to belong to *Origanum*, attracted a lot of interest for their aromatic foliage. *Monarda didyma* and *M. fistulosa* were used extensively in public gardens and landscape parks in Germany in the early 1800s but then fell out of fashion until a new phase of breeding in the early 1900s made them a popular, but never major, border plant. Various attempts at breeding varieties that are immune to mildew have been made, but the fungus has generally adapted rapidly. The late-20th-century interest in native North American flora stimulated greater awareness of the natural diversity of the genus in the wild, which has resulted in a wider gene pool in cultivation.

OPPOSITE *Monarda didyma* by Dutch painter and illustrator Aert Schouman (1710–1792); the style (a single plant in the landscape) evokes that of the previous century. Watercolour? on paper, dated 1753. Courtesy of the Rijksmuseum, Amsterdam.

Muscari Asparagaceae

There are 42 species of *Muscari* (grape hyacinths), all bulbs, found from the Maghreb and Iberian Peninsula across to central Asia. The name derives from the Turkish for the plant. As with many bulbs, there is a distinction between those that are generalists and those that are found naturally in situations, such as dry hillsides or steppes, where a very short growing season is the norm, the latter being more difficult in cultivation. Most *Muscari* species are very much the former, growing in a wide range of habitats, including grassland and open woodland, making them easy in cultivation. Some have an especially long growing season, leafing up in autumn to make the most use of a Mediterranean winter's light. Most have adapted well to cultivated environments, naturalising in orchards and vineyards, as well as in gardens in cooler and wetter climate zones, to the point of becoming almost weedy. Some species are eaten in regional Italian and Greek cuisines; the flavour of the bulbs is apparently bitter, but a process of cooking and then preserving in oil reduces this.

Muscari neglectum was known to German botanists in the 16th century and appears to have been distributed throughout Europe from this period onwards, becoming especially popular during the 18th century but never achieving the high status or price of tulips or hyacinths. Further introductions arrived in Europe over the next few centuries. During the 20th century, colour variants were actively sought out and propagated by the bulb industry—different shades of blue, white (where all pigments are missing), pink (where the blue but not the red pigment is missing), and ice-blue (where the red is missing).

> Symbolic of the good old days of the Rococo; matt colouring is balanced with delicious fragrance.
>
> (Carl Bolle, 19th-century German naturalist, quoted in Krausch 2003)

Narcissus Amaryllidaceae

There are some 70 species (and 27,000 cultivars) in the genus of the quintessential spring flower, whose centre of diversity is the mountains of the Iberian peninsula and the mountains just across the water in the Maghreb. One species, *Narcissus pseudonarcissus*, has a wide distribution in western Europe; *N. poeticus* (pheasant's eyes) and several related species are found across the regions immediately north of the Mediterranean. The heavily fragrant white *N. tazetta* is found further eastwards around the Mediterranean into Iran; long traded on the Silk Road, it got as far as Japan centuries ago and has naturalised there. As with tulips and lilies, there is no sepal/petal distinction; what are commonly called petals are referred to technically as perianth segments. The daffodil cup (i.e., the trumpet), however, is a structure that has evolved independently and is unique to daffodils, which are conventionally categorised into 13 divisions:

- Trumpet (Division 1). Daffodils where the length of the cup is as great or greater than the length of the perianth segments. Each stem has a single flower.
- Large-cupped (Division 2). Daffodils where the length of the cup is more than a third the length of the perianth segments but less than equal to it. It is the largest division, some 45% of cultivars.
- Small-cupped (Division 3). Daffodils where the length of the cup is less than a third the length of the perianth segments. They show the input of genes from *Narcissus poeticus*, so there is a tendency for strongly coloured cups and white or pale perianth segments.

A plate from an early-20th-century book on bulbs, illustrating the different divisions of the genus *Narcissus*. From *Beautiful Bulbous Plants for the Open Air* (Weathers 1905), with coloured plates by Mrs. Philip Hensley.

- Doubles (Division 4). Most of these have a somewhat messy appearance, looking very much the mutants that they are.
- Triandrus (Division 5). Descended from *Narcissus triandrus*, with a distinct bell-shaped cup, and usually with at least two flowers per stem.
- Cyclamineus (Division 6). Heavily influenced by *Narcissus cyclamineus*; trumpets have perianth segments which are slightly bent back, giving them a windswept look. Many are early-flowering and generally smaller-sized.
- Jonquils (Division 7). Descended largely from *Narcissus jonquilla*; typically heavily scented, with narrow leaves, they have traditionally thrived in the American South.
- Tazettas (Division 8). Largely descended from eastern Mediterranean *Narcissus tazetta*; like tulips, they need a summer baking. Hardier plants for northern winters are the Poetaz Hybrids, made between these and Poeticus daffodils.
- Poeticus (Division 9). Derived from *Narcissus poeticus* and related species. The cup is very shallow and often brightly coloured, so the overall impression is of flat white perianth segments with a central ring.
- Bulbocodium (Division 10). Derived from the very distinctive *Narcissus bulbocodium*, with its lampshade-shaped cups and minuscule perianth segments. A relatively new division, with breeding recently underway.
- Split-corona or Split-cupped (Division 11). Daffodils whose cup is split and splayed out. These arouse strong feelings, as many people see them as unnatural mutants.
- Miscellaneous (Division 12). Another division where no one species dominates the genetic make-up. This is an increasingly important group as breeders have now brought every species into the gene pool.
- Natural species, variants, and hybrids (Division 13).

Daffodils reappear faithfully every year, not just in gardens but wherever they may have been dumped decades ago—for these are true clonal perennials. The plants make the most of the long growing season offered by the western Mediterranean climate, where a long cool moist winter is followed by a spring with reasonable soil moisture levels. In the wild they are plants of light woodland, and open, but not unduly exposed, country. They are beneficiaries of traditional agricultural practices, as part of hay meadow or grazed grasslands; as a result, agricultural intensification or the abandonment of hillsides to scrub hits them hard.

There has been little herbal or other practical usage of the plants: the generic name derives from the Greek, *narco* ("becoming numb"), the same root as the word "narcotic," a probable reference to their moderate toxicity. Narcissus was also the beautiful boy in Greek mythology who was punished by the gods for his insensitivity to the wood nymph Echo, who had fallen in love with him, his punishment being to fall in love with the first face he saw, which turned out to be his own as he leant over water to drink; we have the word "narcissistic" as a result. Another Classical legend is that of Proserpine, who stopped one day to pick a daffodil and was kidnapped by Pluto, god of the underworld; eventually a deal was struck and she was allowed back

OPPOSITE An illustration of Small-cupped daffodils, "Four Pretty Narcissi," from Britain's *Amateur Gardening* magazine, c. 1910. The uppermost, 'Persian Orange', and the two below it, 'Dairymaid' and the one on the left, 'Blood Orange', were bred by Engleheart. The lower right-hand bloom is 'Peveril' (unknown breeder, pre-1907). The open character of the flowers—in particular the gaps between the perianth segments—is very characteristic of this early period of daffodil breeding.

to the surface every year. White forms of *Narcissus tazetta* are traditionally used to celebrate Spring Festival in China, while daffodils generally are part of spring festivities throughout Europe and the Middle East, and increasingly globally.

In Medieval Christian art, the flower was used as a symbol of paradise and triumph over death; it is often associated with the Virgin Mary. In the Muslim Middle East it is seen as a symbol of death and planted on graves, because its growth in spring reminds people of the life to come. In Classical Arabic poetry, Poeticus daffodils are seen as having "eyes," and therefore of being the eyes of the garden, as well as symbols of love, longing, and desire. The connection with Wales and St. David's Day (1 March) dates back only to David Lloyd George, who as prime minister in the early 20th century decided that a daffodil would be more attractive in his buttonhole than the traditional leek; they are known in Welsh as "St. Peter's leeks."

White Tazetta daffodils are known from the tombs of ancient Egypt. Occasional mentions are found in Classical texts, and the plants were grown by the Byzantines. *Hortus Eystettensis* (1614) illustrated 40 distinct varieties, including several Jonquil and Tazetta types, two interspecific hybrids, six doubles, and a number of European species. Parkinson mentions 78 varieties.

Daffodils were key to the growing understanding of natural hybridisation and evolution. William Herbert (1778–1847) suspected some daffodils he had found growing wild in France (as well as some garden varieties) were hybrids. He made crosses himself and concluded that some of the wild plants were indeed natural hybrids, which led him to suggest that it should be possible to deliberately make crosses to generate new ornamental varieties. Herbert's revolutionary work encouraged others. Amateur grower Edward Leeds (1802–1877) played a crucial role in early daffodil breeding, as did a dynasty of amateur botanists and breeders, the Backhouses. William Backhouse (1807–1869) worked on colour, initially using Poeticus varieties and moving on to *Narcissus pseudonarcissus*. His son, Robert Ormston Backhouse (1854–1940), along with his wife, Sarah Elizabeth (1857–1921), went on to become the most prolific daffodil breeders of the family. George Engleheart (1851–1936), who worked on daffodils between 1882 and 1923, is often regarded as the man who really launched the modern daffodil, producing some wonderful white varieties and others with bright orange cups.

Percival D. Williams (1865–1935), one of the greatest of all breeders, began his efforts in 1895; his great achievement was the Large-cupped 'Carlton', the most widely sold daffodil of all time and an essential cut flower for an industry that became increasingly important from the early 20th century onwards. Alec Gray (1895–1986) raised 110 new daffodils over 59 years of active breeding, with a focus on miniatures, often using seed from wild plants collected in Spain and Portugal. Among his creations was 'Tête-à-Tête', which by 2006 made up 34% of Dutch bulb production, with 17 million pots sold at auction. The Dutch themselves had become important breeders in the late 19th century but have always been very commercially focussed, with most breeding done by companies rather than individuals.

Grant E. Mitsch (1907–1989) in Oregon was the first really systematic American breeder, from the early 1930s onwards. The first known daffodil show in the United States was run by the Maryland Daffodil Society in

Feral daffodils in Herefordshire, England. *Narcissus pseud-onarcissus* is widespread in western Europe but almost certainly not native in Britain. It is thought to have been introduced to Britain by the Romans 2,000 years ago, and there are now many, very localised, colonies, such as this one, growing in the light woodland habitat typically pre-ferred by the species.

1924, with the Garden Club of Virginia setting up shows 10 years later. This latter group played a major role in promoting the flower, setting up a system of daffodil test gardens in different parts of the state in 1930. Led by the Virginians, interest in daffodils grew, and in 1955 the American Daffodil Society was formed.

The breeding of daffodils has historically been led by the first three divisions, to produce large plants, with flowers suitable for cut flowers as much as garden use. Over time, interest in the smaller divisions has increased, using Cyclamineus for earliness and diminutive size, Jonquils for scent and heat resistance, Tazettas for scent and floriferousness, etc. Breeders, who are now very widely distributed—in eastern Europe and Japan as well as the longer-established ones in the former British colonies—are increasingly focussed on small plants. Some of the most adventurous work is being carried out in California, using a very wide range of wild species. The best breeders are still amateurs and dedicated enthusiasts rather than companies.

Nepeta Lamiaceae

A varied group of herbaceous plants, comprising about 250 species, found across Eurasia and parts of Africa, named for a Classical-era town in Tuscany. Several species contain nepetalactone, a chemical compound that attracts cats; *Nepeta cataria* is the best known of these catmints, as they are popularly known. The species in cultivation tend to reflect the large number which are found in seasonally dry habitats in the Mediterranean and the Middle East. Many, including some in cultivation, are plants of central Asian steppe grasslands. Others (e.g., *Nepeta govaniana* and *N. subsessilis*) favour cooler and moister habitats; these are predominantly the eastern Asian species. All would appear to be long-lived clonal perennials with a competitive character.

The entire genus appears to be chemically interesting, semiochemicals (i.e., compounds which signal to and cause a response in animals) being a particular feature: *Nepeta racemosa* is being trialled for use as an insect repellent in crops, while *N. cataria* has traditionally been used as a treatment for various pests of poultry and is now being trialled as a mosquito repellent. Catmint appears to have very mild psychoactive properties; in traditional herbalism it was used as a sedative and had a reputation for stiffening the courage of the timid. In the Middle East, several species are used to treat depression, among other applications; overcollecting for herbalism is proving a conservation issue in the region.

Nepeta cataria has been grown in herb gardens since the Medieval period, and during the 19th century several further species were introduced and used ornamentally.

> Turneiserus lays down an observation concerning a hangman, that was otherwise gentle and pusillanimous, who never had the courage to behead or hang any one till he had first chewed the root of cat-mint.
>
> (Jacob 1836)

Neither Silva-Tarouca nor Robinson has much to say about them; it would appear that the popularity of the genus came later, largely fed by the English Arts and Crafts school of garden design. *Nepeta ×faassenii*, a sterile hybrid of unknown garden origin in the 18th century, is now one of the most common of all garden perennials in Britain. Since the 1970s it has been joined by an increasing number of other species and hybrids, a popularity partly driven by a growing appreciation of their drought tolerance, longevity, and compact size.

Nymphaea Nymphaeaceae

Nymphaea (from the Greek for "nymph") includes some 40 species, with a global distribution. Waterlilies are among the most primitive of all flowering plants—fossils have been found in Jurassic deposits, and many species are thought to have changed little since. During warm episodes in the earth's history, the genus became very widespread, leading to its wide, but disjunct distribution. Nymphaeas are herbaceous aquatic plants, although some species can adapt to growing terrestrially on damp ground. They normally grow in still, fresh, neutral to alkaline water, with different species adapted to flourish at different depths. Most species produce both subaquatic leaves and the well-known "lily pads," which sit on the water surface; some tropical species produce foliage that stands proud of the water as well. Tropical members of the genus (and their hybrids) may be either day- or night-blooming; the latter are pollinated by moths.

> [M]y favourite is the waterlily. How clean it rises from its slimy bed. How modestly it reposes in the clear pool, an emblem of purity and truth.
>
> (Chou Tun-i, 11th-century Chinese writer, quoted in Swindells 1983)

The waterlily of the Nile, *Nymphaea lotus* (not to be confused with the true lotus, *Nelumbo nucifera*), was of immense symbolic importance to the ancient Egyptians, who made great use of the flower in making wreaths for altars and funerary rights, as well as for secular gifts and decoration. The plants have featured widely in art and ornamentation. All species are potentially edible but are rarely a common food item.

The tropical species exercised a particular fascination for 19th-century gardeners; the first hybrid of these was created by van Houtte's nursery in Belgium. Other early hybrids were developed at the Berlin Botanic Garden. But the breeding of waterlilies thereafter is very much linked to one man, Joseph Bory Latour-Marliac (1830–1911) of the Lot-et-Garonne department of France, who produced the Marliacea Hybrids (*Nymphaea alba* × *N. odorata*) for larger pools; the Laydekeri Hybrids, derived largely from *N. tetragona*, for smaller water bodies; and the Pygmaea Hybrids (*N. tetragona* × *N. mexicana*) for depths of only 30cm. Although he was quite open about his methods (unlike many breeders), he was careful as a businessman to name and release onto the market only sterile hybrids, so no one else could use them in their own breeding. All told, Latour-Marliac produced around 100 varieties during a life dedicated to the plant. More valuable work with these plants was done at the Missouri Botanical Gardens by George Pring from 1915 onwards.

Glendale Flower and Water Gardens

M. I. STOLER, Proprietor

1260 JUSTIN AVE., GLENDALE, CALIFORNIA

Lotus

Water Lilies

Aquatic Plants

Cut Flowers

Fish Food

Scavengers

Fancy Gold Fish

1. Rose Arey—
   $2.00 each; $2.15 Postpaid
2. Zanzibarensis Rosea—
   $1.50 each; $1.65 Postpaid
3. Zanzibarensis Purpurea—
   $2.00 each; $2.15 Postpaid
4. Conqueror—
   $2.50 each, $2.65 Postpaid
5. Marliac Chromatella—
   $1.00 each; $1.15 Postpaid

Special Offer — This entire "Paramount" Collection of 5 Superb Water Lilies (Regular Value $9.00) - - - for $7⁵⁰ PostPaid

An illustration from the catalogue of Glendale Flower and Water Gardens, a water plant nursery that thrived in California in the 1920s. Several tropical hybrids are shown: 'Rose Arey', 'Zanzibarensis Rosea', 'Zanzibarensis Purpurea', 'Conqueror', and 'Marliac Chromatella'. Courtesy of Robert Smaus.

Orchids Orchidaceae

With between 22,000 and 26,000 species, divided among 800 genera, the Orchidaceae is arguably the largest flowering plant family. The genetic flexibility of the orchid genome is illustrated by there being over 100,000 hybrids, many of them intergeneric—more than almost any other family, orchids are able to successfully breed across wide taxonomic boundaries. Modern taxonomy divides orchids into five subfamilies: Apostasioideae, Cypripedioideae, Orchidoideae, Epidendroideae, and Vanilloideae; of these, the first and last are horticulturally almost irrelevant. Cypripedioideae includes the slipper orchids, with their distinctive pouch-like lip, most of which are terrestrial, including the familiar tropical Southeast Asian *Paphiopedilum*, and the temperate northern hemisphere *Cypripedium*. Orchidoideae is terrestrial, and includes the vast majority of temperate zone species, as well as the increasingly familiar South African *Disa*.

But where evolution has truly come out to play is Epidendroideae, with its vast array of epiphytic and lithophytic species, mostly to be found in warm-climate zones. It includes 80% of Orchidaceae species, in approximately 570 genera. Of these, the following are the most widely grown:

- *Cattleya*. The classic orchid, the one that adorns chocolate boxes and any other merchandise seeking to establish its luxurious credentials. All epiphytes of montane South American origin, the closely related *Laelia*, *Brassavola*, and *Sophronitis* were crossed with *Cattleya* early on, with more genera being added over time, to produce one of the most genetically complex groups of ornamental plants. In popular culture this is the ur-orchid, its association with Gilded Age plutocrats having shaped the whole social perception of the family.
- *Cymbidium*. Found from southern China down to northern Australia, the ability of *Cymbidium* to withstand relatively low temperatures has made it particularly versatile. The flowers are valued commercially for their robust texture and longevity.
- *Dendrobium*. Around 1,200 species are found from China and India down to New Zealand and across the Pacific Islands. The most commercially important are the den-phals (dendrobiums with a phalaenopsis-type flower), classically associated with Thailand.
- *Masdevallia*. The ultimate cult orchid genus. Around 500 species are found in the cloud forests of South America. Their vividly coloured and extraordinarily shaped

flowers attract dedicated growers who have to deal with what are often tricky cultural requirements.

- *Miltonia*. A small genus from the coastal montane rainforest of Brazil. It has been extensively hybridised with the closely related *Oncidium* and several others to produce a range of hybrids known as pansy orchids for their flat-faced flowers. Miltonias are among the most amenable to cultivation as houseplants.
- *Odontoglossum* and *Oncidium*. Two large South American genera, most species of which are found in montane regions. They and related genera are currently in a state of taxonomic muddle, as might be guessed from a ferocious level of intergeneric hybridising. The flowers offer an enormous range of shape and colour, irresistible to the breeder.
- *Phalaenopsis*. From southern China down to Queensland, from both lowland and montane rainforests, the moth orchids were minor players until the late 20th century, but extensive hybridising (including much intergeneric) and the development of robust tetraploids have resulted in this genus now being the most popular of all. Modern hybrids seem almost ideally suited to the temperatures, humidity, and light levels of the modern home.

An orchid display at the Royal International Horticultural Exhibition of 1912, which was held at The Temple in London, where RHS exhibitions were held prior to the first Chelsea show in 1913. The photograph was reproduced in *The Horticultural Record*, published in 1914 to celebrate the event. Such displays are still a feature of shows such as Chelsea. Courtesy of RHS Lindley Library.

Orchids are extraordinary, and somehow apart from other flowering plants. The flowers are nearly always instantly recognisable as orchids, while their habit of growth has several features which set them apart, and in some cases are unique. Two types of growth are recognised. In monopodial orchids there is one growth point, with new leaves added at the top and the stem becoming ever longer; the plant is essentially non-clonal, although in some species occasional plantlets are produced. In sympodial species, growth is lateral, usually with a storage structure (the pseudobulb) being replaced every year, with new growth appearing at the "front" end, and old pseudobulbs lining up "behind." Some sympodial species are strongly clonal, forming clumps. Most familiar orchids are sympodial; the moth orchids are distinctive in that they are monopodial. Terrestrial species often have tubers—the word "orchid" is derived from the Greek for "testicle," after the shape of the tubers of some of the terrestrial Mediterranean species; they may be rhizomatous and clonal, forming clumps, or may be essentially non-clonal.

Above all, orchids have a mystique—the product of their often unusual cultivation requirements, the undeniably sexual quality of the flowers, and the late-19th-century orchid craze that swept western Europe, a craze which established them as plants for a wealthy and decadent elite. To botanists there is a mystique too: the incredible range of species, the rarity of many, and, especially, the highly specialised and almost arcane range of pollination mechanisms. Orchids today, however, have lost some of this mystique, ironically because of some of their other special qualities: their resilience (they may be difficult to grow well, but they are also difficult to kill), the prolific ease with which they seem to cross, and the several techniques that may be used to propagate them, which—although laboratory- rather than potting-bench-based—have made it possible to produce plants on a truly industrial scale. Science, with plenty of help from low-cost skilled Asian labour, has now democratised the orchid.

One orchid myth can be firmly quashed. The amazing proliferation of species, their pollination technology, and their total absence from the fossil record convinced many that they evolved in recent geological time. No! A paper in *Nature* in 2007 announced the discovery of fossil orchid pollen, along with a bee, in amber, from the Miocene. This, along with studies of change in DNA over time, seems to indicate that orchids first evolved in the late Cretaceous and diversified and spread around the world after the asteroid impact at the Cretaceous-Paleocene boundary. They are therefore much older

One small autobiographical piece by [Richard] Rorty bears the title "Wild Orchids and Trotsky." In it, Rorty describes how as a youth he ambled around the blooming hillside in northwestern New Jersey, and breathed in the stunning odour of the orchids. Around the same time he discovered a fascinating book at the home of his leftist parents, defending Leon Trotsky against Stalin. This was the origin of the vision that the young Rorty took with him to college: philosophy is there to reconcile the celestial beauty of orchids with Trotsky's dream of justice on earth.

(Habermas 2007)

Disa uniflora (as *D. grandiflora*). A drawing made on Table Mountain, during one of Robert Gordon's expeditions to the Cape Province, sometime between 1777 and 1786. Courtesy of the Rijksmuseum, Amsterdam.

than we thought. New species can evolve very quickly however, partly because of the very close relationships between many orchids and the particular insect species that pollinate them.

Orchids are found in almost every habitat on earth but never the very harsh ones. Perhaps the most clearly unique feature of orchids is their relationship with fungi. Many plants are associated intimately with mycorrhizal fungi in symbiotic relationships, but orchids carry the relationship to an extreme. Orchid seed is so minimal (each capsule releases around a million dust-like seeds) that it consists of little more than a poorly developed embryo and will not germinate unless certain fungi are present. The process can be bypassed in cultivation by germinating the seed in a sterile nutrient solution.

Around 70% are epiphytes (overwhelmingly the subfamily Epidendroideae), with the greatest diversity in the cloud forests of the Andes and the mountains of Central

OUR ORCHID COLLECTOR, MR. LOUIS FORGET, PASSING A VILLAGE IN THE U.S. OF COLUMBIA WITH A CONSIGNMENT OF CATTLEYAS.

"Our orchid collector, Mr Louis Forget, passing a village in the U.S. of Columbia [sic] with a consignment of cattleyas" says the title of this photograph included in the 1919 catalogue of Sanders, one of Britain's leading orchid dealers. Such collecting did enormous damage to South American orchid populations. Today the main threat is habitat destruction driven by coca farming for cocaine production. Courtesy of RHS Lindley Library.

America; the climate typically gives relatively cool nights and constant humidity. In some cases species can be endemic to extremely small areas; in any case, the epiphytic lifestyle offers a plethora of microhabitats. Epiphytes and lithophytes are often adapted to survive seasonal drought; some have CAM photosynthesis; they are also noted for their aerial roots, which have evolved to be able to extract moisture from the atmosphere. A crucial adaptation to seasonal drought is the "pseudobulb," the storage organ prominent on most epiphytes and lithophytes. Orchids which lack this are those from environments, such as equatorial tropical rainforest, where there is no dry season (e.g., the Southeast Asian *Phalaenopsis*).

Terrestrial species are often highly particular about soil conditions, owing to their relationship with mycorrhizal fungi, but tend to be found in woodland glade and edge habitats, scrub, and open grassland, with a strong correlation with lower-fertility habitats. Investigating the orchid-fungus relationship has been made considerably easier by DNA analysis of the fungi, which suggests that in the wild these orchids are highly specific about the fungus species they will team up with—and helps explain why many terrestrial orchids are difficult to grow. The orchid seems to control the relationship very closely, and to benefit, both in terms of carbohydrate and mineral take-up. Indeed, so little does the fungus gain, that it might be more accurate to describe the orchid as parasitic on the fungus; in cases where the fungus is also involved in a mycorrhizal relationship with tree roots, the orchid is indirectly parasitising the tree. Some terrestrial species have taken the relationship with fungi to an extreme, losing the ability to photosynthesise, even living their entire lives underground.

Traditional societies tended to pay orchids little attention, and there is a relatively sparse folklore attached to them. Any cultural associations preindustrial societies had for them were usually to do with either fertility or elegance. Some terrestrial species have been used as food or flavouring, notably *Orchis* species in the Middle East (e.g., in making the drink salep), and of course the familiar vanilla pod is from the striking epiphyte *Vanilla planifolia*. Some species, notably *Dendrobium nobile* in China, have been used in traditional medical practice.

Since their discovery by Western ornamental horticulture, orchids have become very important as commercial cut flowers and have been extensively used for symbolic purposes, e.g., as national flowers or representing cities or regions. They are some of the most important plants semiotically, with their mention in literature nearly always representing luxury and decadence. The flowers do seem to have a strong tendency to evoke female genitalia—during the 19th century, when public prudery and private vice were frequently separated by little more than a door, this must have been an important part of their appeal. *Cattleya* species, with their exotic scents, immense frilled petals, and beckoning symbolic vaginas, were hugely popular in 19th-century Europe, the age of the multi-layered petticoat and crinoline of female attire. Cultural historians would no doubt like to make connections between the French love of cattleyas and the can-can dancers of the Moulin Rouge. In more recent years, the rarity of many species and the criminality associated with their removal from the wild have also fascinated writers (e.g., Susan Orlean's *The Orchid Thief*, which records a real-life story of plant poaching).

> Today, orchid growing is more than just an industry, it is an international business.
>
> (Griesbach 2002)

Warm-climate orchids

The Chinese were the first to cultivate orchids, during the Wei (220–265 AD) and Chin dynasties (265–317 AD); during the succeeding Tang dynasty, the growing of *Cymbidium* species (primarily *C. ensifolium*) became widespread. The earliest orchid-growing manual dates from 1233, the Southern Song dynasty, by which time several other *Cymbidium* species had been taken into cultivation. Buddhist monks had long since taken plants to Japan, with cultivation becoming well established during the Heian.

Awareness of the richness of tropical orchid flora came relatively late to Europeans. The 1818 introduction of *Cattleya labiata* from Rio de Janeiro state in Brazil to England made a big impact; cultivation was initially hampered by a failure to understand its growing requirements—plants were frequently overheated in greenhouses—but by the 1830s, gardeners employed by the nobility in England worked out how to grow the plants successfully. An important breakthrough was the work of Sir Joseph Paxton (1803–1865), architect and horticulturalist, who standardised orchid culture, the key concepts being both good ventilation and humidity. It now became possible for nurseries to start specialising; one of the first was Messrs. Loddiges of London. Importations increased and, crucially, some key works were published. One in particular, James Bateman's *Orchidaceae of Mexico and Guatemala*, was brought out in stages from 1837 to 1843, and at the time was the largest book in print. European orchid fanciers

"The Cattleya House at the Royal Exotic Nursery, King's Road, Chelsea" in the 1880s. The nursery was part of the Veitch business, which dominated late-19th-century British gardening. Illustration from the first volume of *A Manual of Orchidaceous Plants Cultivated Under Glass in Great Britain* by James Veitch and Sons. Courtesy of RHS Lindley Library.

maintained greenhouses with thousands of plants and were prepared to spend small fortunes on obtaining new ones. They were the ultimate status symbol for the nouveau riches of a rapidly industrialising Europe. The appetites of the wealthy were now whetted, and with rapidly growing transport links, several decades of ruthless plunder were unleashed onto the orchid habitats of the world.

Orchid hunting was one of the most shameful spectacles in horticultural history. Whereas most plant collectors were happy with seed and a few plants, the contemporary inability to propagate orchids, meant that plants had to be collected from the wild. Vast areas were ransacked; in some cases, collecting meant that thousands of trees were felled to strip them of their orchids, with some collectors regularly destroying plants they could not remove in order to stop others' getting hold of them. Several species were driven to virtual extinction. Wild collecting all but ended in 1973, when the Convention on International Trade in Endangered Species (CITES) of Wild Fauna and Flora made it illegal (although orchid rarities are still occasionally decimated by determined collectors).

One reason why orchids seemed somehow apart from other plants, and why importation was the only way of obtaining them, was the extreme difficulty of propagation. During the 1890s, however, Joseph Charlesworth, owner of a nursery near Bradford in northern England, began to successfully germinate seeds on nutrient jelly in sterile flasks. This reduced the pressure on wild populations and also enabled hybridisers to stand a good chance of achieving success with the pods of dust-like seed which followed on from their efforts with the pollinators' brush.

Veitch's nursery were pioneers in orchid growing and in hybridising other plants, so it is not surprising that they produced the first orchid hybrid, between *Calanthe triplicata* (as *C. furcata*) and *C. sylvatica* (as *C. masuca*). Flowering for the first time in 1856, it was a botanical and horticultural sensation; John Lindley, Britain's leading botanist and promoter for hybridisation, named it *C. ×dominii* in honour of John Dominy, the nursery foreman who had made the cross. Others hybrids followed, among them *Phalaenopsis ×intermedia*, the first phalaenopsis hybrid, in 1875.

Obsession led to some good "gentleman science," with serious collectors sponsoring the study and classification of orchids, a great many publications, and conferences, the first being held in London in 1885. By this time the mania for growing orchids had spread from Britain to France, Germany, Belgium, and the United States. The early years of the new century saw the pace of hybridisation quicken, along with the cultivation of orchids for the cut flower trade. With Europe exhausted and impoverished by World War I, Americans soon took the lead. A breakthrough was the realisation that cymbidiums could be grown outside in much of California, and by the time the American Orchid Society was founded, in Boston in 1920, pioneer growers had been busy for several decades.

Orchidists suffered considerably again in World War II, with the destruction of the glasshouses at Berlin-Dahlem by Allied bombing and the Philippine National Herbarium by the Japanese army, two particularly serious blows. American growers increasingly favoured Hawaii as a location, as the island's varied climates suited a wide range of species, with *Vanda* a favourite genus for commercial cultivation. Polyploid hybrids emerged in the postwar period; such plants were often more vigorous, easier to grow, and had bigger, fuller, and so more commercially appealing flowers.

In 1960 propagation took another leap forward when a French researcher, George Morel, discovered how to tissue-culture the meristem (growing point) of some orchid genera, enabling the mass production of genetically identical plants. The plants' extraordinary genetic flexibility led to the creation of ever-more-complex crosses, with some new genera composed of crosses of four others. The naming of new hybrid genera developed a surreal quality: the 1988–90 period included the registration of *Nornaham-amotoara*, *Georgeblackara*, *Charlieara*, and *Yonezawaara*. Many hybrids never become commercialised but remain the playthings of hobby breeders.

From being impossible-to-propagate rarities, orchids have now become a globalised industry. Mass production concentrates on *Phalaenopsis* (although many other genera have come down dramatically in price as well). Moth orchids, the ideal plants for the modern home, have become one of the most traded horticultural commodities and one involving complex routes: cultivars may be bred in the United States, tissue-cultured in Japan, grown on in China, and then finished (i.e., grown to retail size) in the Netherlands. The Taiwanese orchid industry in particular has played a major role in the development of this trade. The success of *Phalaenopsis* has put most other orchids in the shade—unlike *Cattleya*, *Cymbidium*, *Odontoglossum*, and most of the other genera which were so popular in the great days of orchid growing, its members are not dependent on the seasons and so are capable of flowering for at least nine months of the year.

Cool-climate orchids

Terrestrial orchids have attracted nothing like the attention of their tropical brethren, although their history in cultivation has likewise been marred with much plunder from the wild. Some *Dactylorhiza* and *Epipactis* species are relatively easy to grow, if slow to propagate, and there is little doubt that a more prominent horticultural future awaits them. It is probably only a lack of investment in researching commercial propagation that is keeping them from the garden centres. *Cypripedium*, being more exotic-looking but also more difficult, has presented an enticing challenge which now looks like it is on the cusp of being met, as several relatively robust hybrids are now available.

And finally, the east Asian *Pleione*, not terrestrial but cool-growing, has transformed temperate zone orchid growing. Plants are relatively easy to grow and to propagate, and a century or so of introductions and hybridising has resulted in a wide pool of popular cool-greenhouse orchids.

Paeonia Paeoniaceae

Named after Paeon, a student of Asclepius, the Greek god of medicine and healing, who became jealous of him, but who was saved from his teacher's anger by Zeus, who turned him into the flower. The genus, the only one in its family, is relatively primitive; it is thought to have evolved early, during the Cretaceous, and then to have undergone rapid evolution in the Pleistocene, as ice ages disrupted the distribution patterns of the original species. There are 30 species, two in western North America, the rest in the Old World: North Africa, the mountains of Europe, and on into central Asia, China, and Japan. Peonies are best understood as falling into four main categories:

- Herbaceous peony hybrids. *Paeonia lactiflora* dominates this gene pool, which also involves several other herbaceous species.
- Tree peonies. A group dominated by selections of *Paeonia suffruticosa*.
- Species peonies. A few species are relatively common in cultivation (e.g., *Paeonia mlokosewitschii*); others are grown only by connoisseurs.
- Itoh peonies. These are hybrids between tree and herbaceous varieties, bred initially in Japan by Toichi Itoh (d. 1956).

Wherever they are found, peonies are often rarities, with localised distributions. They are slow-growing but immensely long-lived plants, taking years to build up root mass; once established they are very resilient to environmental stress and physical damage. Technically, they combine stress-tolerant and competitive strategies: they dominate their immediate environment for the long-term but are not able to extend that domination; a good analogy is yews, which are nothing if not long-term, dominating their immediate environment but without dominating the wider habitat. Peonies in eastern Asia are plants of woodland glades and scrubby thickets; in southeastern Europe, Turkey, and the Caucasus, they are more likely to be found in open, often seasonally dry habitats. It seems highly likely that habitat destruction and exploitation for herbal medicine have massively reduced the wild populations of many species.

Fully mature peonies are the pride of every gardener even if they flower for only a few days. They seem to radiate an elemental power.

(Gerritsen and Oudolf 2000)

Peonies depicted in books compiled by Tachibana Tamotsukuni, 1755. The high status of peonies and their symbolic link with wealth was one of many Japanese cultural borrowings from China. Also shown is a Chinese-style eroded rock and mushrooms—both highly symbolic in Chinese culture. Courtesy of Chiba University Library.

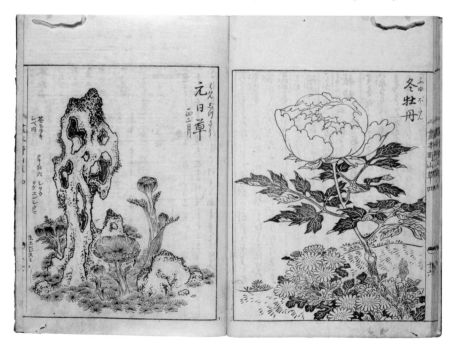

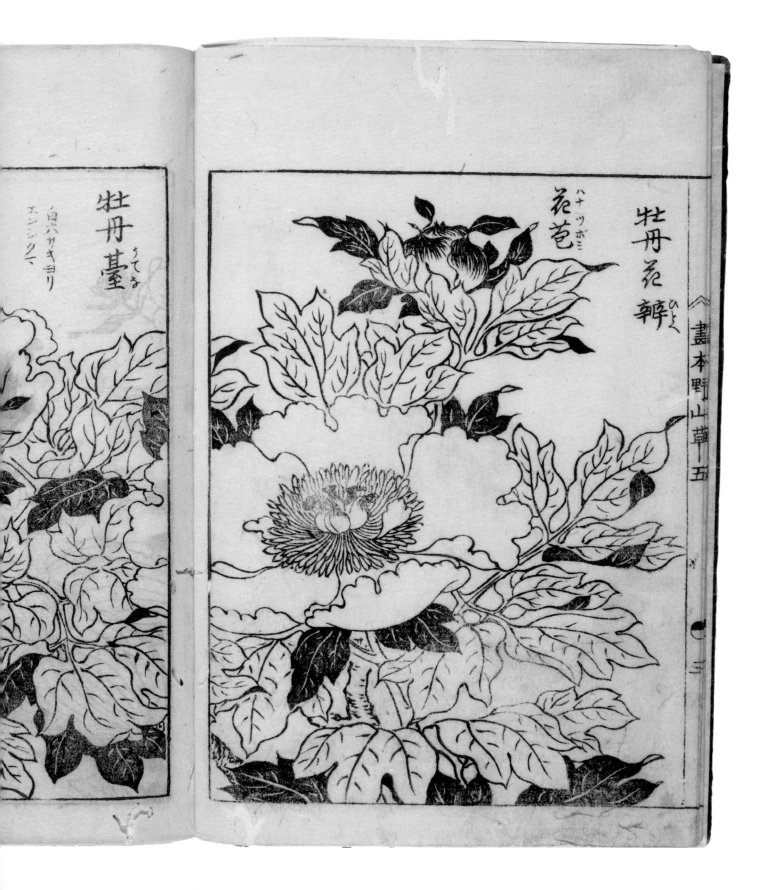

牡丹花瓣（ヒラ）

（ハナツボミ）花苞

牡丹臺（うてな）

ノ白ハサキヨリ
エンジクハ

三

Peony roots have a bitter flavour, which in the ancient world was associated with medicinal value, and extracts from them have been used extensively in both Western and Eastern medical traditions, e.g., for jaundice and kidney problems and to treat a number of neurological conditions. Modern research indicates that the plants are indeed rich in chemicals with a range of effects on the body, but no particular contemporary use stands out. In China, the medical system tends to use them for gynaecological problems: for thousands of years there, the peony has symbolised femininity, as well as wealth and honour. The importance of the peony in Chinese culture is enormous; it is featured in art more than any other flower, although often very stylised. The city of Luoyang in Henan is an established centre for peony growing; a major peony festival is held there every year to promote the region for tourism.

Paeonia mascula and *P. officinalis*, the historic species of European cultivation, were first grown as medicinal herbs; Hildegard von Bingen mentions their efficacy against fevers. By the 16th century, however, they were being grown as ornamentals. *Paeonia lactiflora* arrived from Siberia in the late 18th century, introduced by Peter Pallas, a German traveller in Russia. The plant had, however, been in cultivation in China for centuries, with around 40 cultivars by the Song dynasty; generally these were double, as were most of those which European breeders went on to produce.

Paeonia suffruticosa was cultivated prior to the Tang dynasty, research in China suggesting that it is the result of centuries of development from the wild *P. ostii*, crossed with *P. rockii* and two other, now rare species from Shaanxi Province. It was during the Tang that it really took off as an ornamental, although only among the aristocracy—a concession to the sumptuary laws used extensively by both the Chinese and Japanese to control rank and status. During the Song dynasty, cultivation of the suffruticosas was liberalised, and the scholar-bureaucrats took to the plants extensively, often developing special gardens for them. From around 20 varieties in the Tang, there were 114 by the Song, with doubles particularly highly rated and in a wide variety of colours.

OPPOSITE The front cover of *Kelway's Manual*, 1898, featuring "one of Kelway's new paeonies." Courtesy of Kelways Plants Ltd.

First mentioned in a Heian poem, tree peonies were popular in Japan, too, although they never achieved the popularity they had in China. During the Edo period they were one of the "six excellent flowers of Higo," and part of the culture of Bushido horticulture in the southern island of Kyushu during the 1750s. Some flowers may have been 30cm across. The Higo-shakuyaku Group was particularly notable, flowers being lotus-shaped with prominent golden stamens. During the Meiji era many of the Edo varieties were lost, although a thriving export trade eventually encouraged new breeding.

The first European breeder was Frenchman Nicolas Lemon (b. 1787), who worked with *Paeonia officinalis* and *P. lactiflora*. The latter species showed a wide range of colour variation, and a fragrance far greater than was shown by European species. Lemon was followed by the Comte de Cussy, who was able to import plants from China as well as do his own breeding. Eventually, via another nurseryman breeder, his plants ended up with Lemoine, who, by working with *P. lactiflora* and *P. wittmanniana*, produced a range of hybrids which by all accounts took Europe's gardeners by storm, including 'Sarah Bernhardt', named for one of the great celebrities of the age and still available today; Lemoine also made several adventurous crosses with tree peonies.

Paeonia suffruticosa was introduced to Europe from a von Siebold shipment from Japan in 1844; soon after came *P. rockii*. Both caused major stirs—these were some of the most prized plant introductions of the time. They have never been regarded as that reliable in European climates, however; it is the herbaceous varieties that often survive for years in country gardens.

Kelway's Manual

ONE OF
KELWAY'S
NEW PÆONIES

1898

100 Gold & Silver
Medals Awarded

Langport
Somerset

Seeds Plants & Bulbs

For much of the 18th and 19th centuries, British pottery manufacturers did their best to copy the higher status wares of China (and latterly, Japan), presenting Western consumers with stylised images of Chinese garden landscapes. Peonies—albeit very heavily stylised—are the most regularly depicted flowers. This is a Masons Ironstone bible pattern serving plate, late 19th century.

American interest started early, with production of plants not just for the garden but also for the cut flower trade. Oliver Brand (1844–1921) had accumulated 1,000 cultivars of herbaceous peonies as breeding stock by 1894, and in the 1920s he and his son were the largest producers in the world. The American Peony Society was founded in Detroit in 1903 by commercial growers; among their first actions were setting up a committee to clarify confused naming and establishing a research plot at Cornell University.

During the late 19th and early 20th centuries Kelway and Son were the key British nursery breeding peonies, many of which are still available. Peonies were the jewel in the crown of a very diverse garden plant portfolio that around 1900 employed 400 people and was one of the largest nurseries in the world.

A chemistry professor, A. P. Saunders (1869–1953) of Clinton, New York, took up peony breeding as a hobby and went on to be regarded as the father of the modern peony, thanks to meticulous record keeping and advanced crosses involving as many as four species. He brought in genes for fiery scarlets from *Paeonia peregrina* and *P. tenuifolia* and, most difficult of all, for yellow, from the beautiful "Molly the Witch," *P. mlokosewitschii*. After something of a postwar decline, the entry of Chinese sources into the global market after trade liberalisation in the 1980s greatly widened the gene pool and stimulated increased interest. Peony breeding is once again on the upswing, with production for the cut flower market an important component of the business.

Papaver Papaveraceae

Papaver (from the Latin for the plant) contains around 70 species, widely distributed in the northern hemisphere north of the tropics. It includes some of the most familiar flower icons, known almost universally for the thin satiny texture of their petals, and blooms which always seem so large in comparison to the rest of the plant. Given their strong floral impact and that they are easy to grow, it is surprising, and rather a shame, that so few are in cultivation. The genus has been divided into a number of sections by botanists, but genetic analysis suggests that not all species share a common origin, which in modern terms means that they cannot all be included in the same genus; reclassification and renaming loom ominously.

Plant lifespan varies from annuals, sometimes very short-lived, through biennials to clonal perennials. Annual poppies illustrate very clearly some of the key aspects of the most short-lived pioneer plants: a high proportion of biomass dedicated to flowers—essential to attract pollinators—followed by sturdy seedheads and a large volume of seed, which in some species is able to survive long periods buried in soil, possibly even centuries. Annual species include familiar weeds of arable land (e.g., *Papaver rhoeas*), of the open soils of dry habitats in the Mediterranean and central Asian regions, and of high-altitude or high-latitude environments; Arctic species (e.g., *P. radicatum*) are some of the most northerly of all higher plants. The best-known perennial poppies,

Papaver orientale varieties are visible in the foreground here, a display border at Kelway's nursery from the 1909 edition of *Gardens of Delight*, a promotional booklet that was used to market the company's concepts of ready-made borders (which could be ordered by the foot) and site-specific plantings (for those who could afford it). The company had extensive borders for trialling plant combinations and for impressing visiting customers. Courtesy of Kelways Plants Ltd.

the *P. orientale* types, originate from an area encompassing the Caucasus and parts of central Asia, growing in steppe, mountain grasslands, and scrub. These are true clonal perennials with a lifecycle reminiscent of geophytes, making rapid early growth, flowering, and then going into mid-summer dormancy.

Red poppies appear in many cultures but with varying symbolism, sometimes for death (as eternal sleep) and sometimes for resurrection. *Papaver rhoeas* has an important symbolic role in Britain for its link with World War I, as vast amounts of buried seed germinated in the soil of the battlefields of northern France; paper ones are sold every autumn to raise money for veterans' charities. Widely used as a source of seed for bread and cakes, *P. somniferum* is most famous and infamous as the source of opium. Long used as an analgesic, it appears to have become first widely used as a recreational narcotic in 15th-century China. It was extensively grown in 19th-century England for the production of laudanum; this century also saw the beginnings of today's drug trade when the British government forced the Chinese to accept trade in opium in the notorious Opium Wars. It is still widely grown for medical use, but it also fuels the illegal production of heroin, so providing the income for a plethora of terrorist and criminal groups.

> In Flanders fields the poppies blow
> Between the crosses, row on row,
> That mark our place; and in the sky
> The larks, still bravely singing, fly
> Scarce heard amid the guns below.
>
> (John McCrae, Canadian soldier, May 1915)

Papaver somniferum has been grown for millennia, and in northern Europe since the late Medieval period at least, with some ornamental ones known by the late 16th century. Double forms are thought to have been introduced from the Ottoman Empire around this time. Their very colourful flowers and immensely easy propagation from seed made them a popular cottage garden plant, especially in those of the poor. *Papaver nudicaule* came from Siberia early in the 18th century, one of several species introduced from far northern regions, along with *P. radicatum*; these, and several other short-lived perennials, have been crossed and recrossed to produce the familiar Iceland poppy, usually going under the name of *P. nudicaule* (of hort.). Surprisingly for such delicate flowers, they cut well and have become commercialised as cut flowers.

Annual *Papaver rhoeas* has produced the Shirley poppies, thanks to William Wilks (1843–1923) of Shirley in Surrey, southern England, who collected seed of atypically coloured poppies from the fields and bred them into a colourful seed strain. Wilks was a leading amateur researcher into heredity and played a crucial role in chairing one of the world's first conferences on genetics in 1900.

Papaver orientale came to Europe via Tournefort, who found it in Armenia in 1701. It soon became very popular in German gardens. *Papaver bracteatum* was introduced from Russia in 1817, the two crossing to produce today's hybrids at the end of 19th century—the *P. orientale* of gardens. Perry was instrumental in its initial breeding. In the 20th century the plant was most strongly associated with the great perennial nursery of Countess Helen von Stein Zeppelin in southwestern Germany.

Papaver orientale varieties from Foerster's *Gartenstauden Bilderbuch* (1938), illustrated by Esther Bartning. The caption bemoans the lack of resistance to frost, rain, and storm of the 150-odd cultivars then in commerce. These, however, were recommended as being among the most resilient—clockwise from upper left: 'Colonel Bowles', 'Olympia' (described as the "crown of all"), 'Württembergia', 'Prinzessin Victoria Louise', 'Roland', and an unidentified white. Courtesy of Bettina Jacobi.

Pelargonium Geraniaceae

Any discussion of *Pelargonium* has to start with establishing why, in popular parlance, these plants are called geraniums. Originally all were classified in *Geranium*, which Linnaeus agreed with, despite the opposition of some botanists, who believed the bilateral symmetry of *Pelargonium* flowers clearly set them apart. By the beginning of the 19th century, the distinction had been established, but by this time, it was too late, and the English-speaking world has known them as geraniums ever since. The name comes from the Greek for "stork," after the beaked shape of the seedheads.

OPPOSITE An illustration from *The Flora Magazine*, a British publication of 1867, depicting three *Pelargonium* cultivars: 'Ajax', 'Ocellatum', and 'May Queen'.

The approximately 280 species include shrubs, subshrubs, climbers (or more accurately scramblers), perennials (some with tubers), and annuals. Two species are found in Turkey, eight in Australasia, and 18 in East Africa; of the rest, the vast majority are South African, and most of them in the Cape Province. Of these, 22 are at serious risk of extinction. *Pelargonium* species are generally plants of drier sites—rock outcrops, dry grassland, scrub—and many show various stress-tolerant adaptations, such as aromatic leaves (to deter predators). Some however, although not generally species in cultivation, are shade tolerant or prefer moist sites. Many, particularly those which have become important in cultivation have an almost shrubby habit and illustrate a long-lived competitor habit through a tendency to regenerate from the base. The broad categories of horticultural pelargonium are fairly clearly distinguished:

- *Pelargonium zonale*. The ancestor of the default zonal or garden geraniums, introduced in 1710 and crossed in only a few years with *P. inquinans* to produce a tetraploid hybrid, *P. ×hortorum*, to which gene pool others were subsequently added.
- *Pelargonium peltatum*. The ancestor of the trailing (or indeed climbing) ivy-leaved geraniums, including the Balcon varieties, made famous world over by Alpine houses and the chocolate boxes that picture them.
- *Pelargonium cucullatum*. The ancestor of what the English call regal pelargoniums and the Americans rather charmingly call Martha Washingtons, presumably as the first First Lady grew them.
- Angel pelargoniums. Crosses between *Pelargonium cucullatum* and *P. crispum*, a species with citrus-scented foliage. They have much of the appeal of the regals but are more compact.
- Scented geraniums. A catch-all for species grown for their highly aromatic leaves, as opposed to their (usually rather quiet) flowers.
- Other species. All the others: an assemblage of species grown by enthusiasts (or of interest only to botanists) and some old groups no longer commercially important.

African peoples have used the species for a wide variety of medicinal uses, very few of which got transmitted to Europe with the plants. An exception was *Pelargonium sidoides*, which has been used in a number of herbal remedies for respiratory conditions since the late 19th century, despite charges of quackery by orthodox doctors. The oils of some species are used in aromatherapy.

C.T Rosenberg del & lith.

Printed by C.F. Cheffins London

Pelargoniums.

1. Ajax.   2. Ocellatum.   3. May Queen.

Alongside the vibrant colours of zonal and other flower-power pelargoniums, the less tangible nature of scent must not be forgotten. The 19th century, indeed surely like all the centuries before it, was hugely dominated by scents—most of them bad, often very bad. The powerful scents of many *Pelargonium* species endeared them to all classes, particularly since many of the aromas mimic better-known ones: rose, peppermint, lemon, etc. The late 19th century saw the start of the cultivation of geraniums for aromatic oil production around the French perfumery centre of Grasse. One of their advantages is that they are far more productive of certain aromas than the "real thing," so making growing and harvesting geraniums more cost-effective than doing the same for roses, mint, nutmeg, or whatever. Today, North Africa, India, and China dominate production. Scented pelargoniums have also been used in a variety of dishes and food preparations, mostly sweet.

Initially the Cape was a Dutch colony, so it was Leiden Botanic Garden that received the first plants, during the late 17th century; the Dutch continued to dominate in making introductions of the plants for some decades. Given how easily pelargoniums can be propagated from cuttings, botanical gardens and wealthy collectors soon established a lively trade in the plants. In Britain, Mary, Duchess of Beaufort (1630–1715), played a key role in popularising and distributing plants in these early years.

The zonal was an immensely popular plant by the mid-18th century, when it was known as *Geranium africanum*. British, and then French and German breeders went to work, producing every colour between very dark red and pure white. The early 19th century saw a veritable explosion of interest in the plants and their hybridising. Perhaps more than anything else, they benefited from the recent breakthrough in glasshouse technology—above all, they would have been very tolerant of the rather erratic heating systems of the time. Early breeders seemed to be interested in colour more than anything, which led to some adventurous hybridising; magazine writers encouraged their readers to not just grow them but have a go at cross-pollination; German breeders alone had produced more than 1,000 varieties by 1850. Later, however, interest settled down to focus on those sub–gene pools where continual flowering could be relied upon.

Supremely adaptable and resilient, pelargoniums survived whereas nearly all the other once-fashionable Cape plants had fallen by the wayside. Those who just wanted them for an event could hire them out for parties and even those who had only windowsill space had an opportunity to grow at least one. By the mid-19th century two key developments were clear. One was the value of the plant for the new fashion for summer bedding schemes, especially with the development of dwarf zonal varieties; indeed the plants became the most reliable member of this new art form. The other was the movement down the social scale, as the tough bright red zonal geranium became *the* plant of choice for those too poor to have the space to grow anything else.

If I do live again I would like it to be as a flower—no soul but perfectly beautiful. Perhaps for my sins I shall be made a red geranium!!

(Oscar Wilde, letter to Harry Marillier, November 1885)

That the zonal geranium could survive for so many years made it the nearest thing to a pet for many of the gardening poor; their longevity tended to breed a level of emotional attachment never before seen with a plant. Inevitably they acquired detractors. By the late 19th century the plants were widely seen as parvenu, vulgar, too tied up with industrialisation, mass production, and the crass new chemical dyes; romantic aesthete John Ruskin fulminated against them, as did utopian socialist designer and campaigner William Morris; Thomas Hardy was one of many writers who used them as a symbol of tasteless modernity, and even the flamboyant Oscar Wilde despised them. There were passionate advocates too, such as Charles Dickens, the great writer on the inequities of life in Victorian Britain—he adored them; the Dickens Society even today has a scarlet geranium (sorry, pelargonium) badge.

The simple scarlet geranium also became a partner in the campaign for virtue. Caring for one was seen as morally improving, so paternalistic middle-class campaigners for morality and against poverty (the two struggles tended to go together) launched flower shows to encourage working-class gardeners to exhibit their plants. Youngsters could be kept on the straight and narrow, away from drinking dens or the street corner, if only they had a geranium to look after.

Other pelargoniums had a different social trajectory—regal pelargoniums stayed very much on the posh side of the tracks, while *Pelargonium peltatum* types became most important commercially, for bedding and other large-scale displays. The scented pelargoniums, particularly *P. odoratissimum* and *P. graveolens* (of hort.) became very popular as working-class houseplants in Germany, but less so elsewhere.

The fact that the geranium became the poor person's flower par excellence continued well into the 20th century. The inhabitants of Alpine villages (once terribly poor!) discovered they could keep geraniums over the winter by hanging them up dry over the winter in a storeroom, planting them out in windowboxes again in the spring. The same trick was used from the 1960s onwards by the inhabitants of communist eastern Europe's jerrybuilt apartment blocks, the cellars of which were full of upside-down geraniums in a state of enforced dormancy during the winter. The "classic" Alpine red geranium is 'Schöne von Grenchen' introduced in 1947, bred by the Swiss company Gärtnerei Wullimann.

Painters always loved the simple scarlet of geraniums. The impressionists in particular made them seem almost classless, as they appear in paintings alongside both high-society ladies and streetwalkers.

The latter part of the 20th century saw the ongoing development of pelargoniums, with substantial sums being invested into breeding ever more varieties. However, unlike with many other commercially important plant groups, there has been little attempt to draw other species into the gene pool established early in the 19th century. The existing gene pool has instead been endlessly sifted for small-scale improvement. Leaf shape and colour has received a lot of attention, as have flower shape (doubles and star-shaped) and plant size and form.

La Belle Alliance.

Pelargonium 'La Belle Alliance' from *The Floral World and Garden Guide* (Hibberd 1858). The name commemorates the alliance of European powers who defeated Napoleon at the Battle of Waterloo in 1815.

Penstemon Scrophulariaceae

Penstemon is a genus of around 250 species, with several hundred more subspecies and distinct geographical races. Species vary in size from 10cm to 3m, with every possible position on the gradient from herbaceous to shrub. Most are somewhere in the middle as subshrubs, being only weakly woody. All appear to be potentially long-lived with the ability to regenerate well from the base, but this is often obscured by their being cultivated in regions where they are not fully hardy. Habitats include desert, open woodland, and alpine; they are generally stress-tolerant plants of drier places, able to reach moisture with deeply penetrating roots. The centre of diversity is Utah, with 50 species, with most of the remainder in neighbouring states, stretching down into southern Mexico, and a few, truly herbaceous, species in the eastern states; indeed every U.S. state, including Alaska, can boast at least one species. The high levels of variation suggests a young genus; the family Scrophulariaceae (now largely broken up following DNA analysis) itself is young, dating back only to the lower Miocene. Plantaginaceae is now proposed for *Penstemon*, whose name means"almost a stamen," a reference to a sterile stamen-like structure in the flower. Various species have been used by Native American healers to treat a range of complaints, including skin and stomach problems and sexually transmitted diseases.

The first scientific recording of a penstemon was made in 1748, but most introductions were made in the 1800s by botanists hot on the heels of the gold diggers, explorers, mercenaries, and outlaws who opened up the American West, including the Mexican portion of the region. Taxonomists subsequently squabbled over names, over which species should be included and which put in separate genera—even over the spelling of the name, "pentstemon" being held up as correct in the earlier part of the 20th century. Hybridisation probably started in the 1830s, in France and Britain, the Germans joining a decade later. It is still not clear which species were used in the original hybrids, but evidence suggests *Penstemon campanulatus* and *P. kunthii* as the source of the genes for smaller-flowered varieties. *Penstemon hartwegii*, discovered in high-altitude Mexico by the epic German explorers Humboldt and Bonpland in 1800, is the parent of most of the larger-flowered modern border varieties along with *P. cobaea*, which possibly was brought into the gene pool in the 1860s.

A Royal Horticultural Society trial in 1861 evaluated 78 varieties—these were probably a small proportion of the vast number of hybrids raised; Lemoine named 470. They were popular primarily as bedding plants, but the hardier ones were frequently seen in cottage gardens. The reason for their success was probably that they could be easily propagated as cuttings in the autumn, overwintered with a minimum of heat (even on the windowsill of a humble cottager), and then romp away when planted out in the spring.

> I call them the *Penstemon Ladies* because I once observed a group of Hardy Plant Society members walking round a garden where they were discussing those ravishing flowers.
>
> (Osler 1997)

In Britain, the Scottish firm of John Forbes of Hawick dominated the growing of the plant from 1870 until 1968. A loss of interest after around 1900 was followed by a fashion for smaller-flowered, hardier, and bushier plants; far and away the most successful of these was the scarlet 'Andenken an Friedrich Hahn' (rechristened 'Garnet' in Britain), bred by a Swiss company in 1918.

Besides a considerable duplication of names early on in the plant's history, there was a very high turnover of varieties, with many disappearing from catalogues after only a few years. Hard winters, too, culled many. English nurseryman Ron Sidwell bred some good varieties in the 1960s, which marked the start of a modest revival in interest in the plant. There was little U.S. interest until the formation of the American Penstemon Society in 1946, and even so, interest there has tended to concentrate more on the desert species and their potential in xeriscaping than on the showy hybrids favoured by European gardeners.

Penstemons on the front cover of the Vilmorin-Andrieux catalogue in 1907, the height of the penstemon boom. Courtesy of RHS Lindley Library.

Penstemon richardsonii in the Columbia River Gorge, Oregon, on a dry south-facing slope. The yellow is *Eriogonum compositum*.

Persicaria Polygonaceae

The name for this genus of 100-odd herbaceous plants is derived from the Greek for "peach," a reference to the leaf shape of some species. Many were once classed as *Polygonum*—including the infamously invasive Japanese knotweed (which has moved on once again, now variously *Fallopia japonica*, *Polygonum cuspidatum*, or *Reynoutria japonica*, depending on which way the wind is blowing). *Persicaria* is cosmopolitan but largely temperate; although many species are annuals and very effective as pioneers, those in cultivation are nearly all strongly competitive long-lived clonal perennials, with varying rates and patterns of spread through rhizomes. The ancestors of the garden species are generally to be found in wet meadows, woodland edge, and tall-herb flora, on fertile, moist soils, often in cooler or montane climates.

All persicarias are edible and make good spring vegetables. The flavour, like that of the related rhubarb, is acidic, owing to the presence of oxalic acid. Both leaves and roots are notably astringent, the roots especially so, which makes them useful in herbal medicine for contracting tissues and staunching blood flow; various species are used in many different traditions.

For the most part, plants entered cultivation in the latter part of the 19th century, as interest in less obviously showy perennials began to grow. William Robinson can be partly blamed for popularising Japanese knotweed, but even he recognised that it was easier to plant than to get rid of. It and several other large-leaved species were highly regarded as landscape plants because of their size and fine foliage, smaller species, for trailing over rockeries. There has been little focussed selection or breeding work, but a steady trickle of new variants, mostly of *Persicaria amplexicaulis*, were developed and marketed through the 20th century.

Dino [Pavledis] came up with several recipes including [Japanese] knotweed and shallot jelly, served with Sussex Slipcote cheese on an oatcake and knotweed compote: knotweed with ginger, raspberries, sugar and vanilla. "You wouldn't identify it because it isn't an identifiable flavour. The wonder of it is that you can use it for sweet or savoury," said Olivia Reid from Terre à Terre.

(BBC Sussex 2010)

RIGHT "A sterling plant, now becoming quite common in good gardens," says *Gardening Illustrated* of the "Japan knotweed" in 1881. To be honest, if everyone had kept reading and followed their advice, there would be fewer problems with this former persicaria: "It should be grown as an isolated specimen . . . in turf." Mowing around this invasive plant is one of the best ways of keeping it at bay.

BELOW *Persicaria bistorta* growing wild along a roadside in Herefordshire, England—a good example of natural clonal (and competitive) spread, as plants rarely reproduce by seed. The selection 'Superba' makes a strongly clump-forming, but not aggressive, garden perennial.

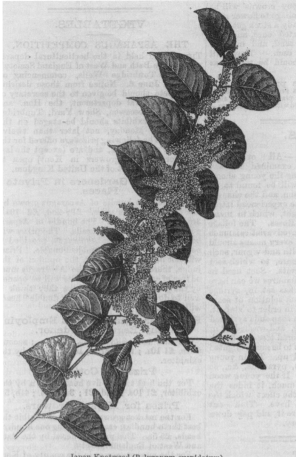

Japan Knotweed (Polygonum cuspidatum).

Petunia Solanaceae

Petunia comprises about 35 South American species, all herbaceous and tending to be short-lived perennials, or annuals. Most species have a pioneer habit, exploiting gaps in existing vegetation in open places. The flowers are popular with butterflies and moths. The name derives from the word for "tobacco" in the Tupí-Guaraní language, whose speakers are widely distributed across South America, and the genus is indeed closely related to *Nicotiana*. The name itself clearly indicates that some species have psychoactive properties. Some tribal peoples in Ecuador are known to smoke the plant, and there has been some interest from seekers after herbal highs.

I'm a lonely little petunia in
an onion patch
And all I do is cry all day.

(Kamano et al. 1946)

    Petunia ×*atkinsiana* is the familiar petunia, a hybrid initially of white *P. axillaris* and purple *P. integrifolia*, both introduced in the early 19th century from Uruguay. The plant spread very quickly across Europe as a summer bedding plant during the 1840s and 1850s, with breeders in England, Germany, France, and Belgium working on the plant. Since these plants are short-lived perennials, it is possible to propagate them from

Perhaps the most popular annual of all time, the stripey petunia encapsulates kitsch in the plant world. This collection of pots was photographed on the tailgate of a camper van parked up in Wales.

A petunia seed strain from Vaughan's 1936 catalogue: Vaughan's Dwarf Small Flowering Mixed.

cuttings, a practice that was widespread in the early days of cultivation; this ensured that the flower colours, and the production of double flowers, were traits preserved from generation to generation. But with the understanding of genetics gained from Mendel's work, known from 1900 onwards, seed production with a reasonably high proportion of desirable outcomes became possible. German breeders led the way, in the 1930s producing the grandiflora types (with big flowers, initially very prone to damage in bad weather) and the multiflora types (smaller flowers in larger numbers on more robust plants).

Japanese breeders also took an interest early in the 20th century, and it was indeed the Sakata company who first stabilised double petunia genetics sufficiently to produce reliable seed. American breeders Charlie Weddle and Claude Hope, founders of the Pan American Seed Company, started to market F$_1$ hybrid seed in 1949 and continued to produce innovative varieties during the 1950s, especially disease-resistant and reliable grandifloras. They produced the first true red—'Comanche', in 1953. Hope produced the first yellow, 'Summer Sun', in 1977. 'Purple Wave', bred by Kirin Brewery in Japan in 1995, was the first of several classes of trailing petunias, which have smaller flowers but because of their ground-covering habit are cheaper to use as bedding.

New classes of petunias are being bred all the time, for this is one of the most commercially important groups of bedding plants. The varieties tend to be grouped in series, with breeders trying to get as many colours out in a particular series format as possible; examples are Cascadia, Conchita, Tumbelina, Fanfare, Madness, Designer, and Potunia (presumably good in pots). Breeders have not only brought genes in from other species but also from the related *Calibrachoa*, itself the result of intensive breeding since the 1990s, to produce ×*Petchoa*, first published in 2007 and since bred by Sakata. Among the most successful of recent developments have been the trailing Surfina petunias, which initially were so difficult to stabilise that producers went back to propagating by cuttings. Surfinas were developed by Suntory, another Japanese brewing and distillery group.

The main ancestor of the many millions of gaudy flowers that adorn windowboxes and summer bedding displays all over the world, *Petunia axillaris* in grassland on the Maldonado Coast of Uruguay. The main grass visible is *Poa lanuginosa*; the blue is *Solanum sisymbriifolium*.

Philadelphus Hydrangeaceae

The genus is named for Ptolemy Philadelphus, an Egyptian king, for no obvious reason. It includes around 60 species, all shrubs, from temperate Eurasia and North America; there are two centres of diversity, southern China and the highlands of Mexico. Plants are known as mock oranges, for their scent (similar to that of citrus flowers); they are also often dubbed syringa: there is confusion with the unrelated lilacs of genus *Syringa*, owing to their being introduced to horticulture at a similar time. They may vary in size, but *Philadelphus* species are all long-lived shrubs, showing continual basal regeneration. Most are found in woodland glades and woodland edge habitats and scrub. Some species occur on dry hillsides, especially the New World species, but most are not notably drought-tolerant. Historically, plants have served as a small-scale source of timber.

Vigour is what these shrubs are known for in gardens, and their "very great plentie" is what Gerard notes they bring to his garden, in the first mention of *Philadelphus* by an English writer in 1597. *Hortus Eystettensis* (1614) illustrates a double. *Philadelphus coronarius*, native to Austria south and eastwards, is the original species in cultivation, with many more being introduced from the mid-18th century onwards, the American ones playing an important role. Lemoine crossed *P. microphyllus*, an almost dwarf species from Mexico, with *P. coronarius* to produce what came to be known as *P. ×lemoinei* and then crossed this with *P. coulteri*, another Mexican with a dark spot at the base of each petal, which continues to mark its descendants.

The delicious scent has always been the main reason for growing mock orange, although not everyone is appreciative—a thread running through commentary on the plant is that some find the smell oppressive; E. A. Bowles thought himself allergic and removed all plants from the garden apart from some variegated ones, which he disbudded. The bulk of the plant, and its relative lack of foliage interest, has made it unfashionable in modern gardens; it has not been taken up by the landscape industry, so the last half century has seen very little breeding.

Phlomis Lamiaceae

The name illustrates a former use: it is derived from the Greek for "flame," as some species were used as lamp wicks in Classical times. There are around 100 species; of interest are those in subsection *Dendrophlomis*, which are subshrubs or very weakly woody, and those in section *Phlomoides*, which are herbaceous. All the former are from around the Mediterranean and western Asia as far as Iran, the latter almost entirely central Asian, although *Phlomis tuberosa* is found as far west as Austria. The Mediterranean subshrubs tend to have a more open habit than much of the maquis flora, and like many tend to deteriorate with age; they are however longer-lived than many. Those in section *Phlomoides* come from a region with a severe continental climate, tending to summer drought; the hairs on the leaves are a good defence against grazing, and so they tend to be a visually prominent part of the central Asian grassland flora. Very few of this latter group are in cultivation; they have enormous potential. There are a number of minor herbal uses. Miller grew several species in the 18th century. Robinson lists 10 but calls them "old-fashioned." A wide range of the shrubby species are now grown, as well as some named selections, but they do seem, on the whole, unjustly neglected.

Phlomis tuberosa (foreground) with *Salvia deserta* (rear) in steppe, Kyrgyzstan.

Phlox Polemoniaceae

There are around 70 species of this genus of herbaceous plants, the name of which derives from the Greek for "flame," after the bright flowers of many. Including both annuals and perennials, *Phlox* species are found throughout North America, as far as central Mexico. Some, not in general cultivation, hail from subarctic regions; one species even makes it over the Bering Strait into eastern Siberia. Phlox vary considerably in their habit and habitat tolerances. None are real wetland plants, but there is a clear gradation, moving from species from damp, fertile soils such as *Phlox maculata* and *P. paniculata*; through woodlanders like *P. divaricata*, which copes with very dry summer conditions; to very tough dwarf rock-dwellers like *P. subulata*. Nearly all those in cultivation are clonal to some extent. As far as the gardener is concerned there are four basic categories, which in reality all segue into each other.

A mixed seed strain of the annual *Phlox drummondii* in Vaughan's 1936 catalogue.

- Taller species and varieties, strongly associated with border planting, mid-summer flowering.
- Sprawling, ground-cover woodland species, early summer flowering.
- Drought-tolerant, low-growing alpine or arid-region species, many from the American West.
- Annuals, e.g., *Phlox drummondii*.

Garden border phlox are derived from *Phlox maculata* and *P. paniculata* and possibly several other species. They are clearly competitors in fertile soil but show considerable variation in vigour and the level of clonality. *Phlox paniculata* as a species tends to form "fairy rings" and so be constantly on the move, a reflection perhaps of an adaptation to an unstable habitat—streamsides and gravel riverbanks. The woodland species vary in the level of shade they tolerate, their preferred pH, and other factors. The alpine species slowly form mats, with many rooting from the stems as they go.

The first seeds of *Phlox divaricata*, *P. paniculata*, and *P. subulata* were sent in 1745 by John Bartram to London, and several species were commercially available by the end of the century. Border phlox and some smaller species were popular by the 1830s, with

A page of phlox varieties in Foerster's *Gartenstauden Bilderbuch* (1938), illustrating a technique occasionally used in German catalogues of the period to show off a wide range of cultivars. Painted by Esther Bartning. Courtesy of Bettina Jacobi.

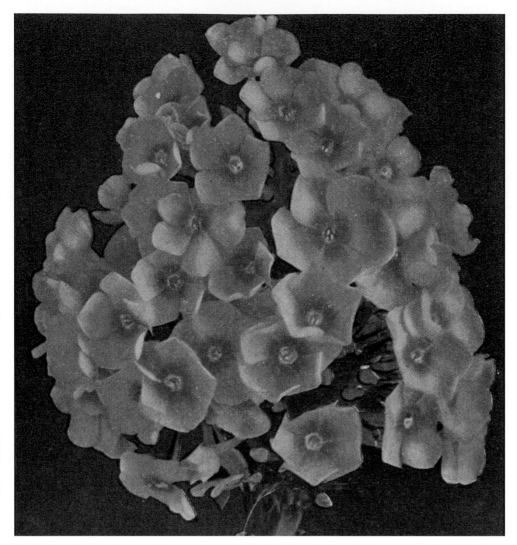

'Daily Sketch', one of many *Phlox paniculata* varieties raised in the 1930s, in Kellogg's 1946 catalogue. It was named for a populist newspaper which ran from 1909 to 1971; Britain has a long tradition of news media sponsoring plants and gardens at the Chelsea Flower Show.

European breeders beginning to export plants. By mid-century American nurserymen were selling "improved" forms, many of them acquired initially from Europe.

The discovery of the annual *Phlox drummondii* in 1834 in Texas was followed by an extraordinarily successful launch into cultivation, the plant soon being praised as "the pride and ornament of our gardens" (Hooker 1840). A few years on, a number of forms were available; there seems to have been a huge range of colour and patterning—pinks, reds, yellows, even dark purples. By 1915 there were 200 named varieties. It is now sadly rather rare.

Phlox paniculata attracted breeders in Belgium, where François Rodigas (1801–1877) of St. Trond produced in 1838 a striped variety that van Houtte, ever the entrepreneur, purchased for 1,200 francs. Foerster was breeding them in the 1890s, and from 1912 to 1927, so was Arends, who crossed *P. paniculata* with *P. divaricata*, producing usefully short hybrids. Early-20th-century Germany went through something of a

Here we go again. Karl Foerster has been quoted time and again, but he was right and so we shall do it one more time: 'Life without phlox is a mistake'.

(Gerritsen and Oudolf 2000)

craze for phlox, with breeding mirroring rapidly changing fashions in colour in a very dynamic garden scene; for example, 1926–27 saw the fashion for red in borders give way to salmon pinks. These are plants which show a great deal of variation and can be bred quickly, so perhaps this flexibility helped drive their popularity. There was surprisingly little interest in Britain until B. H. B. Symons-Jeune (1883–1959) began working on them during the 1940s and 1950s. Some breeding took place in the Soviet Union in the 1950s: the plants flourish in brief but warm Russian summers. Breeding everywhere has continued at a steady pace since the 1960s; there is still a huge amount of scope for the border phlox gene pool.

Interest in rock gardening crystalised in the United States with the formation of the American Rock Garden Society in 1934, about the time that the western desert and mountain species were being collected and trialled in the garden, e.g., *Phlox adsurgens*, *P. bifida*, *P. douglasii*. These seem to cross readily, resulting in many cultivars, e.g., the Millstream Hybrids, all of which arose spontaneously in a Connecticut garden.

Potentilla Rosaceae

The name is derived from the Latin for "powerful"—a reference to the emblematic cinquefoil ("five leaves") of Medieval heraldry, which represented power or potency. Indeed, many European species, known as cinquefoils, typically have five-lobed leaves. And the power continues as far as traditional herbal uses go: in Chinese medicine several species are used to treat diabetes. *Potentilla* comprises around 500 species of annuals and perennials, both herbaceous and woody; these are very much plants of cooler temperate climates across the northern hemisphere, but with outlying populations in montane New Guinea and India's Western Ghats. Some tidying up is needed to clarify the exact content and boundaries of the genus; DNA analysis reveals a close relationship with *Fragaria* (strawberries) and *Alchemilla*, and the woody species (e.g., *P. fruticosa*) are regarded by some as distinct enough to have their own genus, *Dasiphora*.

Potentillas are found in varied terrain (but most frequent in montane environments) and tend towards climate zones with a short growing season. They are generally plants of open situations. Plant behaviour combines competitive and stress-tolerant strategies. Those in cultivation are mostly clump-forming clonal perennials, many of these originating from species that flourish in alpine meadows, where grasses and other plants support their rangy stems. The only subshrub in cultivation is *Potentilla fruticosa* (*Dasiphora fruticosa*). A plant of very wide distribution, it is

> [A] very easy garden plant . . . . Unfortunately, this has made [*Potentilla fruticosa*] ideal for widescale uninspiring abuse on roadsides and traffic islands, especially in the urban sprawl of the 1960s and 1970s.
>
> (Gerritsen and Oudolf 2000)

often found in climates or situations where severe summer drought is combined with winter cold; it can be found in subarctic tundra, in cool deserts, and at high altitudes across both Eurasia and North America.

The story of the herbaceous potentillas is a one of great popularity followed by dramatic decline. Nursery catalogues of the 1860s until the early years of the 20th century could list up to 60 cultivars, probably hybrids of the Himalayan *Potentilla atrosanguinea* and *P. nepalensis*, the latter having very variable flower colours. Lemoine in particular bred many doubles. There are a number of popular survivors from this gene pool, but the range is a shadow of its former self.

Potentilla fruticosa first became popular in the 18th century in English-style landscape parks in Germany; it was also known in Britain at this time. The wide distribution of the plant in Europe meant that a lot of genetic diversity was to hand during the 19th century, even before introductions came in from China and Siberia, which began in the 1820s when Loddiges began receiving Asian-origin seed. Farrer records seeing Tibetan hillsides covered in this species, in every shade imaginable from white through yellow to orange. This diversity was extensively exploited in the 20th century, with a range of hybrids the result.

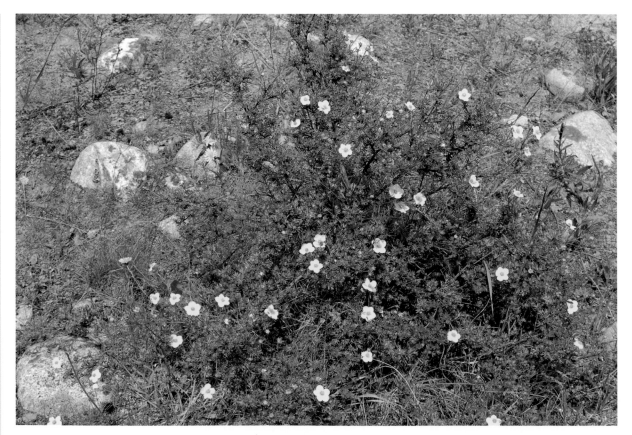

Potentilla fruticosa in the wild in the Ala Archa valley, Kyrgyzstan. The plant is often found in very dry rocky soils.

Primula <small>Primulaceae</small>

Primula is a large and complex genus of at least 400 species of herbaceous perennials, and a few annuals, with 33 in Europe, 20 in North America, one in South America, and the remainder in Asia, mostly in the Sino-Himalayan region. Such a large genus is inevitably divided. Some 40 sections have been recognised but by no means finalised. The name is from the Latin for "first," i.e., the first to flower in spring. Found in a vast range of habitats, a factor species have in common seems to be a preference for "cool"; they can be found on shaded woodland banks, wet meadows, and shaded mountain crevices; in general high altitudes and high latitudes seem to be favoured. A strong characteristic is an ability to grow at low temperatures, which enables them to make a lot of growth in short periods of favourable weather. Most grow and flower rapidly in spring and become either summer dormant or make little growth the rest of the year. Those in common cultivation are all clonal perennials, but not always strongly so. Many species have a tendency to die out over time, as the plants are not particularly persistent, clumps often breaking up or deteriorating after a number of years; a partly pioneer habit is illustrated by their copious seeding. All need high moisture levels but near perfect drainage while in active growth.

The leaves contain an array of chemicals, which makes them unattractive to eat—the downside is that some people suffer contact dermatitis after touching them; some auricula growers can only handle their plants wearing rubber gloves. The resemblance of *Primula veris* (cowslip) flowers to a bunch of keys has led to a range of country names (in both English and German) making associations with the Virgin Mary or St. Peter (who holds the keys to heaven). Cowslip wine is an old English country preparation made from the flowers. It and some other species have been used in European herbal medical traditions for a very wide range of conditions, with little evidence for efficacy.

European primulas

European primulas include the ancestors of the historic and immensely successful polyanthus and the

Six flowers of Alpine Group auriculas shown in a hand-coloured engraving from *Aurikel Flora nach der Natur gemahlt* by F. A. Kannegiesser, published in Leipzig in 1801. Courtesy of RHS Lindley Library.

rather more specialist auricula. Some species common as wild plants—*Primula veris* and *P. vulgaris* (primrose), for example—have benefited from the almost Europe-wide interest in growing native flora.

The familiar polyanthus, sold from filling station forecourts, flower shops, and garden centres, is derived from three familiar species of wide European origin (also extending into Turkey and deep into Russia), *P. veris*, *P. vulgaris*, and *P. elatior* (oxlip), as well as a fourth, *P. juliae*, from the Caucasus. All are members of section *Primula*.

Primula ×pubescens (auricula) is derived from two species native to the mountains of central Europe, *P. auricula* and *P. hirsuta*, these wild ancestors being found in well-drained alpine grassland and rock crevices, often calcareous.

The remainder are rarely grown, although a few alpine species (chiefly *Primula allionii* and *P. marginata*) are grown by alpine enthusiasts; the show-bench sight of a perfect hemisphere of *P. allionii*—almost invisible beneath flowers—is a surreal spectacle.

Auriculas as cultivated plants (i.e., *Primula ×pubescens*) first appear in the historical record in the late 16th century, with Huguenot refugees apparently bringing them to England around this time. They were popular florist flowers, with hundreds of varieties being grown in Britain and in Germany by the late 18th century, both single and double. A decline in the 19th century followed, but the late 20th century saw a revival. The plants became popular with a wider range of non-competitive growers; auricula societies still flourish, and shows are still put on, mostly in the north of England.

Natural hybrids between the three main species in the primrose/polyanthus gene pool are relatively common. Parkinson mentions a red-flowered "primrose" from Turkey, which had been imported for the first time shortly before the publication of his *Theatrum Botanicum* in 1640. A report of the Oxford Botanic Gardens soon after states that white, purple, and blue forms were grown. Polyanthus became very popular during the next century, but the vast majority were completely different from modern forms, being "gold-laced," where the edge of the petals is yellow, the body colour being red, brown, or even near-black. These were extensively grown by the florists. Being easy to propagate and grow, they began to be grown by working-class gardeners in the early 1800s, although usually in pots. Hundreds of varieties were selected, and it is thought that during this period it was the most commonly grown ornamental plant. By the mid-19th century, however, exotic novelties began to displace them, and most cultivars became extinct.

OPPOSITE Four auricula heads appear in this opulent arrangement by Eelke Jelles Eelkema (1788–1839), a renowned Dutch painter of still lifes and landscapes, which also includes daffodils, irises, roses, a hyacinth, a prominent anemone, and, below the main display, an *Ixia* sp., which would have been known from the Dutch colony on the Cape. Oil on canvas, between 1815 and 1839. Courtesy of the Rijksmuseum, Amsterdam.

A display of candelabra primulas featured in the Chelsea Flower Show 1914 catalogue of Bees Nursery of Cheshire, England, who were instrumental in introducing many of the plants following World War I. The company traded until the late 1980s. Courtesy of RHS Lindley Library.

The modern polyanthus begins with Gertrude Jekyll, who acquired a strong yellow one sometime in the late 1870s. It was from this that Blackmore and Langdon, Suttons, and several other British seed companies began breeding. Some exceptionally good work was done by a much smaller-scale operation, by Florence Bellis in Oregon, who, starting with seed from Suttons began to create the famous Barnhaven strains during the 1930s. These notably long-stemmed plants are still popular with connoisseur gardeners, seed being produced by hand-pollination in France. The discovery of *Primula juliae* in the Caucasus in 1900, and its introduction into cultivation soon afterwards by Julia Mlokosiewicz, resulted in the production of short-stemmed polyanthus—very useful for bedding. Hybrids bred between this and *P. vulgaris* at the Czech horticultural centre of Pruhonice and later by Arends circa World War I resulted in a range of multi-coloured compact plants—another significant addition to the gene pool.

Parallel with the production of polyanthus from seed has been the maintenance of named cultivars, propagated by division, the method used by the florists. A few 19th-century varieties are possibly still in cultivation, but the overwhelming majority have been lost. Division has been particularly important for the double varieties, which tend to be sterile. Named cultivars are generally grown only by enthusiasts and produced by specialist nurseries, although some doubles have been commercialised on a larger scale through micropropagation.

Contemporary polyanthus breeding is led by a number of multinational corporations, who use advanced genetics to produce ever brighter and more predictable plants. Millions are grown by large and small scale producers to satisfy a demand for what is basically a disposable plant, to be sold from a vast range of outlets or used in cool-season bedding schemes.

OPPOSITE *Primula sinensis* as the "fruit fly of plant genetics," a diagram by John Innes Horticultural Institution artist Herbert C. Osterstock, made in 1929. John Innes Archives, courtesy of the John Innes Foundation.

A photograph showing Messrs. Rutland and Emarton demonstrating *Primula sinensis* experimental work in the John Innes Library at Merton, Surrey, in the 1930s. John Innes Archives, courtesy of the John Innes Foundation.

L V ℓ

JJ JJ

Jj Jj

jj jj

J.C. Gahrnstock, March 1929.

Asian species

Primula in the Sino-Himalayan region explodes into an exuberant celebration of evolutionary diversity. Section *Proliferae* is the group best established in gardens; these are the "candelabra" species, whose flowers grow in whorls up their stems. Several other species from other sections in this region have become common in gardens, notably *P. sikkimensis* and *P. florindae* from section *Sikkimensis* and *P. denticulata* from *Denticulata*; a great many more are in specialist cultivation, while a few—from high-altitude or other environments that present challenges to cultivation—are in the hands of dedicated alpine growers, notably the intractable beauties of sections *Petiolares* and *Soldanelloides*. A number of species, from a variety of sections, have been grown as short-lived pot plants, some of them with a monocarpic tendency; they are largely of historical interest.

The Asian species are plants of damp but well-drained, and sometimes seasonally dry, habitats. The *Proliferae* and other large species are spring-growing plants of grasslands wet with winter snowmelt, often grazed by yaks or other livestock; primulas are not nice to eat, so tend to get left—they are at a competitive advantage compared to much of the rest of the vegetation. Most Asian primulas are not reliably long-lived, which has plagued their presence in garden history; it has to be assumed that most are pioneers, constantly distributing their seed into new territory. An exception is *Primula florindae*, which seeds vigorously and does seem genuinely long-lived, clonal, and clump-forming. The "difficult" Asian sections are mostly higher-altitude plants, flourishing in gardens in cooler regions, such as Scotland or Scandinavia.

Several species in Chinese cultivation were acquired by Western traders early in the 19th century. Of these, *Primula malacoides* and *P. obconica* became popular in Europe and the United States as short-lived pot plants. Both are distinctly allergenic and only the latter has survived, with its more vicious chemistry bred out of it. A third, *P. sinensis*, one of the greatest mysteries of British horticulture, was brought back to England from China in 1821 by Richard Rawes (1784–1831), the sea caption famous for his camellia imports. Its origin is unclear; quite possibly it had been in cultivation in China for so long that it developed as a separate species. In any case, no wild plant has ever been found, and all the plants in cultivation are descended from this one introduction. It became immensely popular and was developed into a vast number of forms, including ones with fringed petals, divided leaves, and practically every colour except yellow. By the early 20th century it was one of the best understood genetically of all plants as, rather like *Arabidopsis thaliana* today, or the fruit fly in animal genetic research, it played a vital role in the development of genetic science, at least until the late 1930s. *Primula sinensis* was very allergenic, however, and is now, to all intents and purposes, virtually extinct, although it does live on in a hybrid, the rarely seen yellow *P. kewensis*. A number of species from elsewhere in China or from Japan were introduced later in the century; for example, seed of *P. japonica* was collected by Fortune in Japan in 1870, sent home, and shown in flower at the Chelsea Flower Show the very next year.

> There are huge numbers of incredibly lovely primulas from China and the Himalayas which make one feel very greedy. Sadly, though, they are without exception extremely difficult to grow. The well-known candelabra primulas from China and Japan will cope in wet, peaty soil by self-seeding, but we are not very taken with their screaming colours.
>
> (Gerritsen and Oudolf 2000)

Until the opening up of the Tibet/China border region and Nepal (the latter following a treaty with Britain in 1923), the primula heartland was effectively out of bounds. In 1900 there were only around 30 species in cultivation, but by the late 1920s, around 100. A key figure of the 1920s and 1930s was Arthur Bulley of Bees Seeds—he supported George Forrest's collecting and introduced some of the best candelabra species—while Frank Ludlow and George Sherriff introduced some 60 species. Since 1981, joint Sino-British expeditions have added more species or new collections of previously collected ones.

Asian primulas enjoyed something of a craze during the early 20th century, with some fanciers making primula gardens to show off their collections. They were, however, difficult to maintain, and some species were lost. Gradual crossing to form multi-colour hybrid swarms may have looked pretty, but it threatened the genetic integrity of collections. At a few choice country house locations, the plants have effectively naturalised.

Primula sieboldii, one of the species extensively cultivated in Edo period Japan, remains popular. Its Japanese name, *sakuraso*, relates to cherry blossoms and the fact that they bloom at around the same time. Their popularity owes much to the fact that they are easier to grow and much quicker to propagate than many other Japanese cult plants that flower at this time of year, such as *Adonis amurensis* or *Hepatica* species. The shogun Tokugawa was initially instrumental in popularising them, and during the last two decades of the 18th century over 300 varieties were bred, in a range of colours and with striped and fringed petals. Associations of sakuraso lovers were set up all over the country, with competitions an important feature of their activities. Societies continued to form throughout the 20th century, and over 300 cultivars are still extant, including many of Edo origin. Outside Japan, the plants have enjoyed some popularity in North America and Europe, with many new varieties raised, including doubles.

Single and double polyanthus varieties from *Beautiful Garden Flowers for Town and Country* (Weathers 1904), illustration by John Allen. By this time, these more colourful varieties had largely displaced the gold-laced polyanthus that had been so popular in the 19th century.

Prunus Rosaceae

Prunus (Greek for "plum") is a woody genus of around 430 species, found throughout the temperate zones of the northern hemisphere. Including some key sources of fruit and nuts—almonds, cherries, peaches, plums, apricots—the trees often combine beautiful flowers with productivity. Fossils of well-preserved *Prunus* flowers found in Washington State date back to the early Eocene (around 50 million years ago); the genus is thought to have evolved some 10 to 15 million years earlier, during the Paleocene. Traditional taxonomy and DNA evidence divide the genus into a number of subgenera. From the point of view of horticulture, though, there is a similarity to the many species of *Prunus*, with a couple of exceptions. Nearly all are small to medium-sized trees, with spring flowers, potentially edible fruit, and a tendency to be short-lived, often suffering similar diseases. The exceptions are the occasional small suckering species like *P. tenella*, and the very distinctive evergreen laurels, which most people do not realise are so closely related to familiar fruits.

BELOW A print by the great master of the Edo period woodblock print, Katsushika Hokusai (1760–1849), made between 1800 and 1805, showing Fujiyama through a veil of mist and cherry blossom. Courtesy of the Rijksmuseum, Amsterdam.

Evergreen species

Prunus lusitanica (Portuguese laurel) and *P. laurocerasus* (cherry laurel) occur at opposite ends of Europe: the Atlantic Islands and Iberian Peninsula and the Black Sea region, respectively. This indicates they are relicts of the great swathe of evergreen forest that clothed much of Europe in the Cenozoic—that they grow so enthusiastically in Britain may be taken as a confirmation of this (*P. laurocerasus*, especially). American *P. caroliniana* is very similar. All are strongly competitive and long-lived, regenerating from the base far better than most *Prunus* species.

Prunus laurocerasus foliage has an almond flavour, which was made use of in 18th-century cookery. It was also used to flavour brandy; that "note," however, is hydrocyanic acid, which in too large a quantity poisons the imbiber. It was even used for murder; in a celebrated case, Sir Theodosius Boughton was poisoned with it by his brother-in-law in 1780. The "smell of bitter almonds" even entered popular detective fiction as a sign of cyanide poisoning. The fruit and seed also contain cyanide-related toxins. Confusion with the true laurel (*Laurus nobilis*) has been common since its introduction; whereas Roman victors were crowned with *L. nobilis*, *P. laurocerasus* has tended to be used for this purpose since the 18th century, e.g., for decorating British mail coaches after military victories (largely over the French).

Prunus laurocerasus appears in cultivation in the early 1600s, possibly introduced by

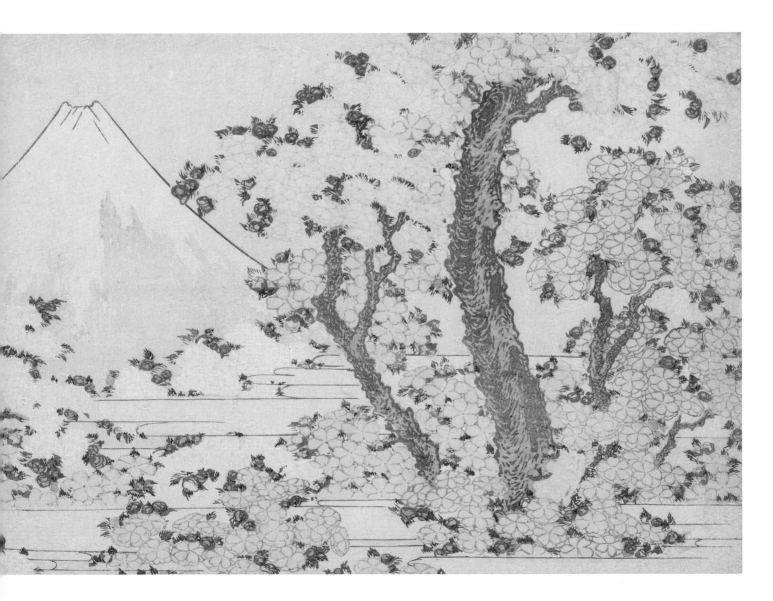

Clusius, who would have obtained it from Ottoman domains. As a highly prized ever-green, during what we now know as the Little Ice Age, the first plants grown were given protection during the winter. By the latter part of the century they were regarded as hardy enough to be grown outside, usually kept trimmed and topiarised, although the large leaf size does not lend itself to the latter use. The landscape movement liberated it to be used on a large scale in woodland, a usage that the Victorians took to with alac-rity—and a legacy many British gardeners and landscape managers continue to live with, and often curse.

The more shapely and slower-growing *Prunus lusitanica*, imported from Portugal in 1719 by pioneering hybridist Thomas Fairchild (1667–1729), was after some cautious experimentation also found to be hardy. Its berries are popular with birds, and at times, have been so with humans, at least in the 18th century, as a basis for jams, and accord-ing to Evelyn, for wine making.

Deciduous species

Deciduous *Prunus* taxa offer a bewilderingly rich and lovely array of trees, grown for their flowers, fruit, and autumn colour, with sometimes an overlap for culinary use. They tend to be pioneer species and therefore short-lived, typically found growing in gaps in woodland, woodland edge, and environments where the larger-growing and longer-lived trees of the woodland canopy cannot dominate, such as steep slopes. Like many pioneer species, they have benefited greatly from the breaking up of historic mature forest by human activity.

Fruit are often eaten by local animal populations. Commercial cherries are pro-duced from several different species or hybrids between them, and there is every rea-son to suppose that other *Prunus* species may yet be domesticated for cropping. It can be assumed that all are probably good for making jam, at least. Various tree parts have been used for herbal medicine down the ages, the bark of the black cherry (*P. serotina*), for example, being particularly valued by Native Americans for its sedative properties. Cherry of various species is deservedly popular as a high-quality wood for turning, fine furniture, musical instruments, and other high-value items, for it is dense and richly coloured. It is also highly regarded as a wood for smoking food. Timbers of other *Prunus* taxa are of similarly high quality but are rarely available in quantity.

The culinary uses of cherries, plums, peaches, and apricots are well known, but it is worth pointing out to Western readers the importance of *Prunus mume* (*ume* in Japa-nese) in the Far East, as a source of fruit for making a very wide range of pickles, alco-holic drinks, and other food products. Japan's umeboshi—plums pickled in shiso (*Perilla frutescens*) and salt—is one of the world's great foods. *Prunus mume* has a major role in Far Eastern artistic culture too, being regarded as another of the Three Friends of Winter (along with bamboo and pine) in classical Chinese art, symbolising virtue and perseverance.

The cultural role of cherry blossom (*sakura*) in Japan is well appreciated internation-ally and scarcely needs to be introduced. The flowers are seen as representing tran-sience, one of the most important values in Japanese culture. Of the many associations

A woman's *haori* (a jacket to be worn outside a kimono) with "plum blossom" (i.e., *Prunus mume*) and bamboo, both of which have been key Japanese design elements since the Heian period. Made anonymously between 1920 and 1940. Silk and gold foil. Courtesy of the Rijksmuseum, Amsterdam.

with the ephemeral beauty of cherry blossom, World War II kamikaze pilots must be mentioned—they too were known as sakura.

Nine species of cherries are native to Japan, with another two introduced from China in the early Edo. Although some 400-odd cultivars were available during the Edo period, the vast majority of cherry trees in Japan today are one variety—*Prunus* ×*yedoensis* 'Somei-yoshino'.

The first recorded plantings were in the 7th century, but cherries played second fiddle to *Prunus mume*, which had associations with Chinese culture and so was valued more highly by the elite. Sakura were appreciated more during the Heian period, and a number of games and party events celebrating the tree's burst of flower were held at the imperial court. Varieties from this time, including colour variants, weeping forms, and double flowers, exist to this day. A key role during this era was played by a temple at Yoshino, not far from Kyoto. Yoshino became a centre for the Shugendo, a mystical sect who blended traditional Japanese beliefs with Buddhism. In their search for contact with *kami* (spirits), Shugendo devotees became pioneer mountaineers. They also started to grow and breed flowering cherries; as an offering to the kami Zaō Gongen, they planted thousands in the valley below the Yoshino temple. Their successors continued, so that today there are around 30,000 trees, flowering in succession, with those on the lower slopes blooming first.

In the Kamakura period, which started in 1185, the eastern regional culture met the Kyoto culture—whose sakura varieties mixed with *Prunus speciosa* in the Kanto region. Many good doubles were the result, the best plants usually being presented to temples. Grafting was also perfected around this time, which helped secure the future of good varieties. During the 15th century, the flower became firmly identified with the Japanese nation.

In the early Edo, hanami parties (originally linking sakura blossom to rituals for a good harvest in the pre-Heian era) were popular throughout the population, with books on the best places to hold them being published. *Prunus pseudocerasus* and *P. campanulata* were introduced from China at this time and joined the gene pool. Singles (not only for cherries but for other plants as well, notably camellias) were preferred; this makes the Japanese almost the only culture who have not always favoured double flowers.

The breeding of sakura reached its zenith in the first four decades of the 19th century, and it was during this time that 'Somei-yoshino' appeared. *Prunus ×yedoensis* is itself a plant of somewhat mysterious origins, a hybrid between two species found in Japan, Korea, and China. 'Somei-yoshino' was popular as it is easy to grow and dramatic: all the flowers come out at once and fall at once. In the widespread social dislocations that followed the 1868 Meiji Restoration, many sakura varieties were nearly lost, and it was only the hard work of a small number of enthusiasts, who collected all the varieties they could, that stopped many disappearing. It was from these collections that, a few decades later, botanists and plant collectors from the United States and Europe were able to start propagating and importing plants.

> Our country is the country of cherry blossom, when people speak of "the flowers" they mean to say cherry blossom. That our country is the cherry blossom country has the same precious meaning as the peony has for Chinese Loyang or the crab apple for Szechuan.
>
> (Keisan Osen, 15th-century Japanese Zen priest, quoted in Kashioka and Ogisu 1997)

A trickle of Japanese cherry imports to the West become a flood in the early years of the 20th century, following the campaigning efforts of a number of Americans who had become enamoured of Japanese culture. Some importations came directly from Japanese nurseries, others through E. H. Wilson and other botanical explorers who travelled widely in Japan. In Britain, plant hunter Collingwood "Cherry" Ingram (1880–1981) had become almost obsessed with them and on a 1926 trip to Japan was able to make his own collections.

Cherry varieties go through phases of popularity: the very bright pink double-flowered 'Kanzan', for example, was planted in vast quantities in newly built suburban areas of London in the 1930s; the very upright 'Amanogawa' was a must in the 1960s British garden; the paler varieties, such as the subtle yellow-green of 'Ukon', are preferred today.

Usually described as a plum, although apparently closer to the apricot, *Prunus mume* is the first significant tree to flower in much of China and Japan after the winter, in late February and early March. There are presently over 300 cultivars in China, with flower colours including white, pink, red, purple, and light green; the tree's history there could date back some 3,000 years. It is thought that *P. mume* was introduced in the Nara period to Japan, where the tree subsequently received a boost from "The Flying Plum," a story about a tree that flew to follow its owner, written by the popular writer Sugawara no Michizane (845–903). Another 200 cultivars were developed during the Edo period. During the Meiji, whereas many other traditional plants lost their popularity, *P. mume* remained a favourite; bonsai was very fashionable at this time, with *P. mume* being a particularly frequent and desirable subject.

A very large number of other *Prunus* species are in cultivation, including ornamental varieties of species normally grown for fruit production, most notably the peach, *P. persica*. Many of these are of eastern Asian origin and were introduced to the West during the late-19th- and early-20th-century period of intensive Asian plant hunting. *Prunus cerasifera* 'Pissardii', a purple-leaved form, was named for its discoverer, a Mr. Pissard, French gardener to the Shah of Persia, in the 1880s; it is a variety of Myrobalan plum, which is put to a variety uses in Iranian cuisine. It has become one of the most common ornamental garden prunus across Europe, with a number of selections available. *Prunus glandulosa* is a short Chinese species long cultivated throughout the Far East with several double forms; now rarely seen, it was popular in the 19th century as a spring flower—young plants would be potted up and forced for early colour. *Prunus subhirtella*, thought to be of Japanese hybrid origin, has become one of the most widely planted, very early-flowering cherries.

Pulmonaria Boraginaceae

There are around 16 species of these herbaceous perennials, known as lungworts in English, after a former herbal usage; indeed, the name derives from the Latin for "lung." Plants are native to cool temperate regions in Eurasia as far east as western China. Pulmonarias are very much woodland perennials, long-lived and slowly clump-forming, although some species exhibit guerrilla spread, too. They have an ability to grow at low temperatures, which gives them an advantage in deciduous forest alongside spring-growing geophytes; like them, they may be summer dormant if conditions are too dry for continued growth. They seem to have a high tolerance of the competition of tree and shrub roots.

The spotted appearance of the leaves of *Pulmonaria officinalis* and many other species led Medieval European herbalists to prescribe the plant for the treatment of lung complaints, because of the supposed likeness between the mottled leaf and the appearance of a lung—an example of correlative thinking once elevated to a principle of medicine, the Doctrine of Signatures. The genus as a whole seems to have made a seamless transition from the Medieval herbal garden to the ornamental border, with other species being introduced into cultivation gradually from the 16th century onwards. *Pulmonaria saccharata* appears in *Hortus Eystettensis* (1614), and cultivar selection started in the late 19th century; *P. angustifolia* 'Mawson's Variety' is one of the oldest. The 20th century saw a trickle of new cultivars; the pace quickened considerably from the 1970s onwards, with specialist nurserypeople like Beth Chatto and Elizabeth Strangman making selections based on good flower colour and foliage patterning.

Pyracantha Rosaceae

Closely related to *Cotoneaster*, the 10 shrubby species of *Pyracantha* (Greek for "fire thorn") are found in a belt from southeastern Europe across to southern China, usually as components of scrub and woodland understorey. They are perhaps most frequent on dry plateaus, on both calcareous and acidic rocks. Birds eat the berries, but different varieties appear to mature at different times, so some will go quickly and others last all winter. Reds seem to go first, orange and yellow later. The berries can be used for making jam.

Pyracantha coccinea, from southern Europe, was the first to be brought into cultivation, at some point in the early 17th century. It was promoted for hedging, for its bright berries as much as its dense habit, John Evelyn (1664) even suggesting that it be added to country hedges—"how beautiful and sweet would the Environs of our Fields be!" Its use as a wall shrub has deep roots—it was being promoted for this purpose by writers in the late 18th century; in some areas this may have been quite a common practice. Asian species started to arrive during the 19th century, initially to Britain and to France, with the inevitable hybridisation to produce today's wide colour range.

Ranunculus Ranunculaceae

Ranunculus is an entirely herbaceous genus of about 600 species. The name derives from the Latin for "little frog," a reference to the wetland habitat of many European species. The genus is old, probably dating back to the Cretaceous, which accounts for its global dispersal. There are parallels with the related *Anemone*, in that the genus is found in some very varied habitats. Overwhelmingly though they are plants of moist habitats, many aquatic, such as the crowfoots. Ranunculus, like primulas, are able to grow at low temperatures and are consequently very much flowers of spring.

Despite being ubiquitous as wild plants in much of Eurasia, few species are important in cultivation, with the exception of one that has been prominent historically: *Ranunculus asiaticus*, one of several Middle Eastern spring ephemerals, has a historical career with an odd equivalence to *Anemone pavonina*. Wild buttercups (e.g., *R. acris*) have a rich folklore but have only been occasionally cultivated, indeed they can be notorious weeds (especially *R. repens*). A double form of *R. aconitifolius* and various natural variants of lesser celandine (formerly *R. ficaria*, now *Ficaria verna*) have been cultivated since the late Medieval period but remained minority interests.

Clusius was one of the first to get hold of something much more interesting from the Ottoman lands—double forms of *Ranunculus asiaticus*. After its introduction to Europe in the late 16th century it was rapidly propagated and became a mainstay of flower painting, florists' competitions, and of the Baroque garden. Two centuries on, it was even suggested that there were more varieties of this than of any other flower, an English catalogue of 1792 listing nearly 800 named varieties. Interest in Britain declined soon after, but double forms of *R. asiaticus* remained popular in central Europe until the mid-19th century, when decline set in there, too. By the 1990s plants could almost have been written up as commercially near extinction, except that interest by the Dutch nursery industry has since revived them, as spectacular but short-lived pot plants.

> They are the greatest ornament of our flower gardens and the pride of the florists.
>
> (Christian August Breiter, 19th-century German gardener, quoted in Krausch 2007)

Rhododendron Ericaceae

With a name derived from the Greek for "rose" and "tree," *Rhododendron* has given the gardening world one of its greatest resources. Many might add "if you have the right soil"—for like most Ericaceae, rhododendrons (and their mycorrhizal partners) depend upon the right soil chemistry to thrive. Distributed across North America and Eurasia, and then round through Southeast Asia to the tip of Australia's Queensland, this is a genus that ranges from mountain tropical rainforest to the Arctic Circle. There are two spectacular centres of diversity: the eastern end of the foothills of the Himalaya and the surrounding region, and the mountain forests of the Malay Archipelago (including New Guinea). The latter region is the home of section *Vireya*, which contains around a third of the approximately 1,000 species. The distinctive pollen of ericaceous plants is first found fossilised in rocks at the very end of the Cretaceous era. The spread of rhododendrons benefited from periods of mild climate, spreading across North America and Eurasia, their range then being much reduced as global climates became colder and/or drier. The Sino-Himalayan region offered a refuge, where they proliferated; and the species of section *Vireya* spread into the mountains of the islands between mainland Asia and Australia, further proliferating. As with all Ericaceae, every species is woody, from true trees (*R. arboreum*, to 20m) down to subshrubs; *R. calostrotum* subsp. *keleticum*, is the smallest, a mat-former from Tibet. Breeders have so far produced over 28,000 cultivars; these are listed in the International Rhododendron Registry maintained by the Royal Horticultural Society.

As might be expected, the classification of *Rhododendron* has caused endless controversy—indeed, the subject could make a long (and possibly rather tedious) book of its own. The first botanist to tackle the genus after the great wave of introductions from Asia was Isaac Bayley Balfour (1853–1922). His creation of 43 series and subseries was intended as a stop-gap measure, but like many such measures, its life proved very much longer. Over time, a more rigorous system based on a hierarchy of subgenus, section, subsection, and species, was developed. In 2004, a major revision, based on genetic and phytochemical data, was proposed:

- *Rhododendron*. The "lepidote" species, with minute scales on the backs of the leaves.
- *Hymenanthes*. The elepidotes (i.e., without scales) and the deciduous azaleas.
- *Azaleastrum*. Evergreen azaleas.
- *Choniastrum* and *Therorhodion*. Both contain obscure but (needless to say) charming oddities.

This is a good start, but taking the classification any further rapidly enters the territory of the serious rhododendron fancier. For the perplexed gardener or garden visitor, the following is offered as a pragmatic (and slightly tongue-in-cheek) overview:

- Deciduous azaleas. Commercially dominated by a series of groups of hybrids generally known by the nursery or location where they were first bred, these are descended from woodland understorey species. Connoisseurs appreciate some delightful early-flowering species, too.
- Evergreen azaleas. Defined as azaleas botanically, these tend to be compact, and the very large number of (often exuberantly colourful) cultivars and hybrids are likewise defined by their place of origin.

- Woodland rhododendrons. The default. The overwhelming majority of species (and hybrids derived from them) grow into plants a few metres high with dark evergreen leaves. These include both North American and Asian species, which naturally grow as a loose understorey in deciduous forest. In the open, they form more compact plants.
- Dwarf rhododendrons. These usefully compact plants originate from moorland and mountain environments, with many coming from higher-altitude areas, or areas cleared by deforestation, in China, Japan, and Korea.
- Large-leaved rhododendrons. Hailing from the humid lower Himalayan foothills, a number of species in subsections *Falconera* and *Grandia* have exotically large leaves, making them look very different from the "average rhodie"; they are notably less hardy, too.
- *Maddenii* rhododendrons. Likewise originating from the lower and warmer foothills, these have the most swooningly exotic of scents but invariably gawky growth habits.
- *Vireya*. These are nearly all epiphytes, not hardy, and form leggy plants, but with the most exquisite flowers in an extraordinary range of shapes and colours.
- Pot azaleas. The ones you buy your mother at Christmas (and which invariably die two weeks later), these are hybrids of *Rhododendron simsii*, an evergreen azalea.

The larger rhododendrons are naturally woodland understorey plants, or in the case of the largest, components of woodland. They thrive well in open conditions as well, which gives them an adaptability, but where they are unwanted, as with the proposed *Rhododendron ×superponticum* in parts of the British Isles, this can be a problem. The smaller species are tough stress-tolerant competitors and, like other ericaceous subshrubs, very long-lived (unlike many subshrubs). Some survive extreme levels of exposure to cold. Not only the vireyas are epiphytic, but some other smaller species in humid Himalayan forest environments are too; there are indeed many warm-climate epiphytes in the wider Ericaceae.

Rhododendrons, and all Ericaceae generally, have a symbiotic relationship with mycorrhizal fungi, which enables the plants to survive in acidic soils where there are very low levels of the key nutrients: nitrogen and phosphorus. The fungi scavenge these elements from a wide range of sources, including dead plant matter (i.e., they can function partly as saprophytes). The mycorrhiza also help protect the plants from high levels of heavy metals and other potentially toxic elements, such as copper, lead, and arsenic. Iron plays an important part in the complex chemistry of nutrient absorption; in calcareous soils, however, iron tends to be insoluble and therefore unavailable to plants—hence the inability of many Ericaceae to grow on alkaline soils. There are exceptions, as some

A variety of flowers in a Greek vase, painted as an Allegory of Spring. *Rhododendron ponticum* is prominent, as is *Kalmia latifolia* (immediately above the top of the vase)—the latter, an eastern North American rhododendron relative, was widely grown in the 18th century but has been displaced as a shrub for acid soils by rhododendrons in European gardens. There are also roses, a sprig of orange blossom, and what could be an amaryllis at the top. Oil on canvas, by Georgius Jacobus Johannes van Os (1782–1861), a member of a renowned family of Dutch artists. Oil on canvas, dated 1817. Courtesy of the Rijksmuseum, Amsterdam.

One of the oldest U.S. public gardens, having opened its gates in 1870, Magnolia Plantation in South Carolina has long been famous for its rhododendrons and azaleas. It has undoubtedly been a trailblazer for a particular use of these plants to provide spectacular colour combinations. This photograph was taken by Frances Benjamin Johnston in 1928. Courtesy of the Library of Congress.

species (e.g., *Rhododendron hirsutum* in the European Alps) are able to grow on limestone; they are thought to form associations with mycorrhiza that are able to scavenge iron at very low concentrations.

Remarkably little traditional use or mythology has attached itself to the genus. The toxicity of some species, and honey made from their nectar, has been well attested at various times in history. Classical sources relate several instances where armies have been decimated by the eating of poisonous honey, survivors reporting hallucinations.

Linnaeus originally created a separate genus, *Azalea*, for deciduous species, long since moved into *Rhododendron*. Several American species made their way into English gardens during the 17th century. *Rhododendron ponticum* was discovered by Tournefort in what is now Turkey and introduced to gardens in the first years of the 18th century, with later introductions of the species coming from southern Spain. A century later, it was being grown in large quantities for forcing by the nurseries around London. The 18th and early 19th centuries saw the American species introduced, and the beginning of a trickle from Asia, including *R. arboreum*, but these seemed only to flourish in a few sheltered gardens.

The first British rhododendron hybrid is recorded in 1814, between a Turkish form of *Rhododendron ponticum* and the American *R. periclymenoides*. Others followed quickly, for the genus is what gardeners, with a dash of prurience, like to call "promiscuous"; seedlings popping up around the borders were often dug up and transplanted to nursery beds in the hope that something better than the parent might result.

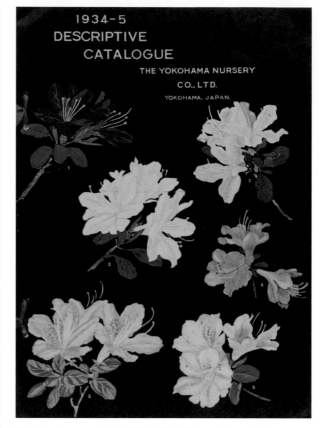

The front cover of the 1934–35 edition of the catalogue of the Yokohama Nursery Co. This decade was the last of what had been around 50 years of export-led prosperity for the nurseries of the area. War meant their destruction; it also resulted in the destruction of a lot of Japanese ephemera in the West—these catalogues are now very rare. Courtesy of RHS Lindley Library.

Rhododendrons provide a good case study of the importance of interspecific hybridisation in horticulture. Pioneer growers in the 19th century found themselves in a frustrating situation—the most sumptuous flowers were produced on plants with somewhat tricky cultural requirements, but the hardiest and most robust plants were never going to win any of the prizes at a flower show. The beginnings of a solution appeared in May 1826, when flower heads of *Rhododendron arboreum* were sent down to southern England to supply pollen to a plant which itself was a cross between *R. ponticum* and the very hardy American *R. catawbiense*. The success of the result, 'Altaclarense', started off a boom in rhododendron hybridising that has never finished. Apparently some 1,800 seedlings were raised from the cross, and widely

OPPOSITE *Rhododendron viscosum*, the earliest azalea species to be introduced to Europe from North America, in the early 17th century. Painting by M. de Gijselaar, dated 27 June 1831. Courtesy of the Rijksmuseum, Amsterdam.

distributed. But the geographical remoteness of the rhododendron centres of diversity meant that the inevitable great flood of introductions had to wait until the very beginning of the 20th century, with fresh material from notable collectors, including Wilson, Forrest, Rock, Kingdon-Ward, and Ludlow and Sheriff. Forrest alone made 5,375 collections of the genus, covering around 260 species.

Joseph Dalton Hooker's (1811–1907) expedition to the Himalayas in 1847–51 brought back 43 species, including some of the very finest. His form of *Rhododendron griffithianum* was much used in hybridisation, one result of which was the 1901 'Loderi'. This was such a magnificent and successful cross that it has been remade over 30 times; large and slow to develop, Loderi Hybrids are not, however, garden centre material. The rhododendron that did democratise the genus was a descendant, 'Pink Pearl', which brought together two hybrids that included Himalayan *R. arboreum* and *R. griffithianum*. Other immensely influential species in developing garden hybrids have been the ferociously scarlet *R. griersonianum* and the dwarf *R. yakushimanum*.

Modern breeding aims at compact growth, extended flowering season, and tolerance of lime. The Inkarho series of lime-tolerant rootstock has taken a good 20 years to develop. Its story began when a self-seeded rhododendron was spotted growing in a quarry with a known calcareous soil and bedrock. This then was crossed with *Rhododendron* 'Cunningham's White'. The resulting progeny of some 20,000 seedlings were grown on in Germany by a consortium of about 20 large nurseries; the chosen one, 'Inkarho', was then mass-produced as a rootstock onto which other established cultivars could be grafted.

Deciduous azaleas

The deliciously scented yellow *Rhododendron luteum*, whose nectar was responsible for the aforementioned ancient honey poisonings, was brought into Western cultivation by Tournefort. It and a number of American species were then bred into a range of highly-scented hybrids by (among others) P. Mortier, a baker in the Belgian city of Ghent, in the 1810s. English nurserymen then worked on these to produce further hybrids. Only around 10% of these original Ghent Hybrids are still with us, however. The 1830s saw

Dutch nurseries work on Japanese *R. molle*, to produce the Mollis Hybrids. Work by the British nursery Waterers and later the Rothschild family, at Exbury in southern England, resulted in the Exbury Hybrids, noted for their rich colouring; much of their gene pool came from *R. occidentale*, a western American species.

Evergreen azaleas

Rhododendron simsii is behind the indoor azaleas sold at Christmas and usually dead within a few weeks of purchase (for which growers are partly responsible, as they tend to heavily root-prune the plants in order to cram them into their pots). Originally introduced to Holland in 1680, *R. simsii* disappeared from Western cultivation until a reintroduction in the early 19th century. Belgian growers have always dominated the market for these intentionally one-season throw-away plants, the Ghent-based van Houtte establishing an early lead—they had 93 cultivars by 1839. There was much German breeding, too, with Dresden being a centre for production for the rest of the century.

Japanese azaleas

In 1918 E. H. Wilson had selected 50 of the best azaleas he could find near the Japanese city of Kurume and sent them to the Arnold Arboretum, from where Dutch and German breeders obtained plants for their own breeding programmes, including Arends, who added the compact, high-altitude *Rhododendron mucronatum* to the gene pool, so creating the very hardy Arendsii azaleas.

What Wilson took away, however, was the result of many centuries of Japanese selection work. *Rhododendron simsii* has been cultivated in China since the early 9th century. Quite when it was introduced to Japan is unclear. Japan itself has another 51 species, with the first taken into gardens in the Heian period, although it was not until the Edo that they really became popular. During this time the demand for housing grew in the capital city of Edo, as regional aristocrats and their vast retinues had to spend alternate years there, attending the court of the shogun; small gardens made the selection of compact varieties essential, or, failing that, small-leaved varieties which took clipping well.

Vireya rhododendrons are now almost unknown to gardeners in the temperate zone. Left to right: *Rhododendron adinophyllum*, from Indonesia; *R. zoelleri*, from New Guinea; and *R. radians*, also Indonesian but—unusually—terrestrial rather than epiphytic.

Rhododendron obtusum and *R. indicum* proved very suitable, as they will still flower profusely if cut before mid-June. It became something of a fashion for lower-class samurai who needed to supplement their incomes to grow and sell the plants. People lower down the social hierarchy who did not have gardens grew them in pots.

It was an early-19th-century lord, Genzo Sakamoto, in Kurume, who discovered how to grow the tiny seeds on moss and developed the first crosses of *Rhododendron obtusum*, *R. kaempferi*, and other species, which are today known as the Kurume Hybrids.

Vireyas

In 1845 the first vireyas were introduced into cultivation in Britain by Thomas Lobb (1817–1894), who was working for Veitch's; the company began hybridising them in the 1890s. They appealed to wealthy gardeners who could grow them under glass and have them in flower over a much longer season than the hardy species. From around six original species more than 500 hybrids were raised by the nursery, including a number of double-flowered "balsamaeflorum" cultivars, the latter unfortunately since lost. Of the rest, only a handful remain in cultivation. They became practically extinct outside botanic gardens after World War I.

By the early 1970s, interest in vireyas picked up, particularly in Australia, New Zealand, and the warmer parts of the United States, with several collectors distributing seed freely to enthusiasts as well as to botanic gardens. Enthusiasts and nurseries started to make crosses, and the group is once again well established as a garden plant in suitable climate zones.

Happily, the fabulous vireyas are enjoying something of a revival in regions with the right subtropical climates: *Rhododendron* 'Highland Arabesque' (left), *R.* 'Kiandra' (below), and *R.* 'Strawberry Parfait' (right)—Australian-bred hybrids all.

Rodgersia Saxifragaceae

Named after 19th-century U.S. naval commander John Rodgers (who collected the type specimen, *Rodgersia podophylla*, from Hokkaido, Japan), the genus includes five herbaceous species, all from eastern Asia. These plants are very similar to the closely related *Astilbe*; *Astilboides tabularis* was once regarded as a member, too. The growth habit is competitive, and they can spread strongly (more so than most astilbes), although it may take several years for rodgersias to establish before they start moving. Like astilbes, they flourish best in streamside and tall-herb flora habitats, where their roots can grow in nutrient-rich and oxygenated water; they can also be found in wet woods. There is some use of the plants in Chinese herbal medicine.

Despite being discovered by Westerners only in the mid to late 19th century, introduction to cultivation was rapid. They arrived at a time when wealthy gardeners were turning their attention away from glasshouses to their extensive acres and to plants that could be grown as understorey to collections of exotic trees and shrubs; in particular, rodgersias flourished as bog garden and lakeside plants. Plants introduced to Pruhonice Park (in what is now the Czech Republic) by Silva-Tarouca in the 1910s are still flourishing. There was little selection work until the 1950s when growers, mostly German, began to make selections, based on flower colour and leaf colour and shape.

Rosa Rosaceae

The rose is *the* garden plant of both Western Christian and of Islamic civilisation, a reflection of a long history of cultivation and a remarkable tolerance and resilience: although lovers of deep and fertile, well-irrigated soils, roses will survive on much less and thrive in climates with fiercely hot summers. The approximately 110 wild rose species are found across the temperate zones of Eurasia, northwestern Africa, and North America, with the greatest diversity in central and eastern Asia. The irony is that in eastern Asia, the region with the most species, they have never been particularly important as garden plants; *Rosa rugosa*, a notably tough species of coastal habitats there, has become something of an invasive alien in parts of northwestern Europe and the U.S. Northeast. "How to kill multiflora roses" is also a frequent Internet search.

> This love is the rose that blooms forever.
>
> (Rumi 1207–1273)

The oldest rose fossil, found in Colorado, dates to 35 million years ago; rates of DNA mutation, however, suggest roses are much older, possibly Cretaceous in origin. The species exhibit a continuum from free-standing shrubs to climbers; sharp thorns (technically these are prickles, outgrowths of an outer membrane—true thorns are modified stems) line the stems, which attach themselves to surrounding vegetation, allowing them to climb. All roses, whether shrubs or climbers, are long-lived. They are found in a wide range of habitats, including sand dunes and desert oases—where deep root systems get them access to moisture. They are characteristic of many transitional habitats, such as woodland edge and scrub. Some of the self-supporting species spread strongly through running roots, a good adaptation to

OPPOSITE Heavily double roses and a few forget-me-nots, painted by Gerardina Jacoba van de Sande Bakhuyzen (1826–1895) between 1850 and 1880. Oil on canvas. Courtesy of the Rijksmuseum, Amsterdam.

constantly changing conditions or shifting sand. The climbers, however, never seem to do this. Some of the climbing species from southern China and neighbouring countries can grow into substantial lianas, as large as the trees that give them support.

A limited number of wild species have been bred over the centuries into a bewildering array of hybrids. The flower shape of wild roses, and the fruit, known as a hip, is very uniform across the genus, but cultivated plants show a wide diversity in the arrangement of petals and the form of the plant. Most varieties are double, and these do not usually produce hips: the petals are so tightly packed, they do not provide access for insects.

There are four subgenera, of which one, *Rosa*, divided into 11 sections, is the source of all the cultivated rose gene pool. Several attempts at bringing order to the profusion of rose species and hybrids have been made. For gardeners, the most meaningful divisions reflect the relationship between founder species and the garden hybrids descended from them. In the historically more recent divisions, however, the sub-genepools may be extremely complex. Only a summary of basic classification will be given here; the three overarching categories in general use by the nursery catalogues are as follows:

- Species roses. The wild species and their cultivars and primary hybrids are the single-flowering, freely hip-bearing, often waywardly vigorous and thorny roses favoured by those wanting the "natural look" (and have both the space to grow them and a pair of strong gloves). A range of sections group varieties linked to one genetically dominant species (e.g., section *Pimpinellifolia*); others cluster related species and all selections from them or hybrids between them (e.g., section *Synstylae*, the Climbing roses).
- Old roses. Varieties with a more complex genetic heritage, bred before 1870. Sections are generally defined by the dominance of a particular species in the gene pool (e.g., Gallica, Alba, Bourbon). Many, but by no means all, have densely packed petals and rich scents.
- Modern roses. Post-1870. Sections here tend to define a characteristic and a history of breeding (e.g., Hybrid Tea, Floribunda) rather than the heritage of a particular species.

"Shrub rose" is a rather vague catch-all term for generally robust roses, which need little maintenance, and flower only once a year; it has been used to cover both species and some old and modern roses. "Climbers" and "ramblers" are two more, rather vague categories. Climbers can, or can be made to, climb; ramblers are lianas—larger, more vigorous, and flower only once a year.

Capturing the scent of the rose has occupied people for millennia, with notable success achieved by the Islamic cultures of western Asia. Rose perfume is made from attar of roses, a mixture of volatile oils produced through the steam distillation of rose petals. Rose water, an aqueous extract, has a wide range of cosmetic and culinary applications, while jams and herbal teas are also made from the petals. The hips, which have some of the highest concentrations of vitamin C in the plant world, are eaten by birds, usually early in the winter (finches also eat the seeds), and are occasionally used in food, usually after processing into a syrup. Kazanlak in Bulgaria has been a centre for rose-based products since the Ottoman period; its annual Rose Festival in the first week of June is a major event.

Roses have a strong link with Islam, particularly with its expression in Persian culture. Through the Mughal Empire this love of roses was transmitted to northern India, where today a range of rose-scented desserts, sweets, and drinks are popular. In the mystic Islamic Sufi tradition the rose is often used as an image of the devotee's search for divine love; rose petals can be bought at many Indian and Pakistani shrines to throw

The opulent display of roses on pergolas or arbours became popular in the early 20th century as climbing varieties with either a long or remontant flowering season became increasingly available. This is a Frances Benjamin Johnston photograph taken at The Fens, a property in East Hampton, New York. Courtesy of the Library of Congress.

onto the tombs of Sufi saints. The rose has been very widely used as a symbol in Christianity as well, the five petals signifying the five wounds of Christ, the bloom itself Mary.

The rose became the national symbol of England in the late 15th century, after a damaging civil war, in which one side had a red rose symbol and the other a white; following the Wars of the Roses, the victor, King Henry VII, combined the two in the Tudor Rose, a symbol of unity. The United States made the rose a national symbol in 1986. It is also frequently used by European socialist or social democratic parties. But primarily the rose has come to symbolise love, the single red Hybrid Tea in particular. Cashing in on this, the cut flower industry grows vast quantities, alongside other colours for less emotionally charged end-uses. Hybrid Teas have tended to dominate commercial floristry, although a minority industry now produces English roses for cutting.

Parkinson had 30 rose varieties in cultivation in 1629; by 1799 around 90 were known. One of the biggest rose collections of all time belonged to Napoleon's Empress Joséphine (1763–1814), who by her death had gathered every type known in her garden at Malmaison. The conquering armies of her husband sent back any interesting roses they discovered, and even hostilities between Britain and France were suspended on occasion to allow for the transportation of roses. Joséphine's patronage encouraged a flurry of breeding across Europe, and it is thought that her head gardener, André Dupont, may even have made some of the earliest attempts at artificial cross-pollination. With the introduction of Asian species, and new hybrids, the number of roses in cultivation began to rise exponentially, with one French catalogue (Desportes) listing over 2,500 varieties in 1829.

Gallica roses are descended from *Rosa gallica*, native to Europe and western Asia, thought to have been brought into cultivation first in Persia in the 12th century BC. Cultivated forms were distributed throughout Europe, possibly by the Romans, and certainly by Medieval monks. More are thought to have arrived from the Middle East with returning Crusaders, including a form, var. *officinalis*, whose petals retained a strong colour and scent when dried and which had astringent properties; this apothecary's rose, as it was known, became popular for medicinal use.

Rosa moschata is supposed to have originated in the Himalayas but is unknown as a wild plant; its strong scent and ability to repeat flower made it useful as the mechanics of breeding began to be understood. A hybrid between it and *R. gallica*, *R. ×damascena* (i.e., of Damascus), is the ancestor of the Damask roses. Recent DNA analysis has shown that the genes of a third species, *R. fedtschenkoana*, are found in some Damasks; this is from the mountain ranges of central Asia and must have arrived via the Silk Route. Damask varieties were introduced to northern Europe in the 16th century.

Alba rose (*Rosa ×alba*), of uncertain European origin, appeared in Classical times; genetic analysis strongly suggests *R. gallica* and *R. canina* (Dog rose) group origin. The latter is a complex of species common across Europe; their genetics are uniquely convoluted.

A Gallica hybrid, *Rosa ×centifolia*, is also regarded as something of an ancestral plant. Thought to be of Dutch origin, the richly scented plant

> We grow roses in all our fifty States. We find roses throughout our art, music, and literature. We decorate our celebrations and parades with roses. Most of all, we present roses to those we love, and we lavish them on our altars, our civil shrines, and the final resting places of our honoured dead.
>
> (Ronald Reagan, Proclamation 5574, declaring the rose the U.S. national flower, 20 November 1986)

OPPOSITE Although the majority of new roses in the late 19th century were bred and introduced by specialist nurseryman-breeders, others were still able to make an impact. 'Jubilee', launched in the Peter Henderson and Co. catalogue of 1897, was bred by E. H. Walsh, an estate gardener in Massachusetts. A pure red Hybrid Perpetual, its description alone took up an entire page in the catalogue. As can be seen, in bud it has the characteristic point of the Hybrid Tea class. Peter Henderson and Co. was a New York–based seed company and nursery.

PETER HENDERSON & CO'S
NEW
GOLD MEDAL ROSE
"JUBILEE"

COPYRIGHT 1897 BY PETER HENDERSON & CO.

has become commercially important as a source of rose oil, and a number of varieties and lineages have developed through somatic mutation. Moss roses, named for the fuzzy aromatic growth on the stems, are largely descended from *R. ×centifolia*.

The lineages just described were the extent of the Western rose gene pool until the very end of the 18th century, when roses from China began to appear. These brought an important new characteristic—repeat (remontant) flowering. One of the first beneficiaries was the group known as Bourbons, which originated on Île Bourbon (a French island colony in the Indian Ocean, now Réunion), where two cultivars were grown as hedging: 'Parsons' Pink China' and the red 'Tous-les-Mois', a Damask with a recurrent-flowering habit. A natural cross occurred in the early 19th century, and the plants were seen by a botanist who sent seeds to France, where breeders crossed it with Damask and Gallica varieties to produce the first group of roses which flowered, if not perpetually, at least off and on through the summer.

The Bourbons, along with a number of other groups that showed a repeat-flowering habit, were intensely bred in the early 19th century, to produce the Hybrid Perpetuals. At around the same time the Noisettes originated in South Carolina, when John Champneys crossed 'Parsons' Pink China' with *Rosa moschata*; the resulting seedlings were handed onto a French immigrant, Philippe Noisette (1773–1835), who sent plants and seed back to a brother in France.

Importations of roses from China were frequent during the 19th century; often they arrived on cargo ships bearing tea, hence, the roses bred from them were dubbed Tea roses. Many had a touch of yellow, were well scented, and (unlike the flat blooms of traditional European roses) featured a slight point in the centre of the flower. In 1867, a rose breeder in the south of France, Jean-Baptiste Guillot fils (1827–1893), discovered a particularly good flower in a batch of seedlings; this 'La France' (1867) was the first of the group that has dominated rose growing and breeding ever since—the Hybrid Teas.

It was Englishman Henry Bennett (1823–1890) who really developed the Hybrid Teas as a class, crossing Tea roses with the hardier Hybrid Perpetuals. The systematic Bennett, perhaps because he was a farmer and cattle breeder, kept thorough records. He was also good at marketing his varieties. Meanwhile, once-blooming roses were rapidly discarded in France, leading breeder Victor Verdier (1803–1893) jettisoning all his once-blooming varieties in 1838 in favour of those that flowered all summer.

Rosa foetida, a yellow species from Persia, had been known in Europe since the time of Clusius but was not successfully exploited until some expertise in hybridisation was gained. The French Joseph Pernet-Ducher (1859–1928), regarded by many as the greatest rose breeder of all time, created the first yellow roses. His Hybrid Tea 'Soleil d'Or' of 1900 was the first "real" yellow, as opposed to the wishy-washy colours of previous attempts; it has been blamed for the introduction of a susceptibility to black spot, as *R. foetida* seems to have a genetic susceptibility to the disease (the disease was not unknown before, however).

Two east Asian species have been immensely valuable as a source of genes, *Rosa chinensis* and *R. multiflora*. The latter, with a sprawling habit and multiple heads of small flowers, introduced some very different characteristics into the expanding rose gene pool. Known as the "seven sisters rose" in Chinese, for the range of different colours visible in a partially open head, it made a major contribution, although not for some decades after its first, early-19th-century introduction. Multiflora genes, combined largely with *R. moschata*, enabled English clergyman Joseph Pemberton (1852–1926) to produce a new range of varieties, profuse and long-flowering and fragrant—the Hybrid Musks. One of his former gardeners, J. A. Bentall, carried on his work after his death producing several classic and still widely planted hybrids during the 1930s.

The Multiflora varieties were the source of a group of late-19th-century hybrids, the Polyantha roses, but also the Dwarf Polyanthas, which proved immensely useful for tidy turn-of-the-century bedding schemes—this was the apex of the formal garden style in Europe. Multiflora-derived vartieties turned out to be relatively hardy, a boon to growers at more northerly latitudes. A Danish nurseryman, Dines Poulsen, and later his brother Svend, used Multiflora, along with Chinensis and Moschata genes, to produce a new class of informal free-flowering varieties, the Hybrid Polyanthas (renamed the Floribundas in the 1950s) from the early years of the 20th century, and they have played second fiddle to the Hybrid Teas in the domination of commercial rose production ever since. The Poulsen company produced the Parade series of miniature roses designed for growing in pots in 1981, and in 1984 the Patiohit series; they have continued to innovate, very much using Polyantha genes, producing ranges of roses tailor-made for a range of situations, from motorway edges to windowboxes.

Some of the biggest rose species are the lianas of the warmer parts of eastern Asia. *Rosa gigantea* from eastern India and *R. wichurana* (memorial rose) from southern Japan, introduced in the 1860s, proved not hardy enough for many European growers, but their genes went into some very good climbers and ramblers. The Noisettes joined with Gigantea and Chinensis genes during the last decades of the 19th century, yielding many good plants—and the pergola, a classic garden feature of the early 20th century, was a perfect place to show them off. Modern climbing roses tend not to be true climbers, i.e., they do not send up long questing stems but are more like lax shrubs, which climb if given a suitable support. Such plants are much more suitable for most private

The "pointy" Hybrid Teas were a special offer in Kellogg's 1946 catalogue. Clockwise from top: 'Étoile de Hollande', 'Pres. Herbert Hoover', 'McGredy's Ivory', 'Pink Dawn', 'McGredy's Yellow'.

gardens than the older large climbers. In the later 20th century, the shaping of roses as landscape plants, simultaneously covering the ground and providing a long season of colour, has been one of the most significant developments, very often using genes from climbing varieties.

Rosa rugosa has been the source of several good selections since its 19th-century introduction. Recent decades have seen it being increasingly used in breeding, especially in Canada. Earlier, however, it has also made a contribution to the Kordes roses, bred by the Kordes company in Germany—tough, disease-resistant, hardy, low-maintenance plants and successful in a remarkably wide range of climates.

Rose breeding from the 1950s onwards was increasingly scientific and ruthlessly commercial. Colours, especially among Hybrid Teas, got stronger and more strident, and breeding for fragrance was a lesser priority. Rose breeding also became very international. Modern varieties may include an enormous amount of complexity—the 'Queen Elizabeth' rose, introduced by veteran breeder and grower Harry Wheatcroft (1898–1977) to celebrate the coronation of Queen Elizabeth II in 1953, was of American origin, bred by Walter E. Lammerts (1904–1996), known in the United States as the father of scientific rose-breeding. Wheatcroft researched its background and found that it was the result of no less than 39 different varieties. Ironically, Lammerts was a creationist.

A portrait of the saviour of the "old-fashioned roses," Graham Stuart Thomas, by Valerie Finnis. Courtesy of RHS Lindley Library.

Inevitably a reaction set in, as many of the older varieties were disappearing. Graham Stuart Thomas made it his mission to collect and therefore save as many of the old varieties as he could; as gardens advisor to Britain's National Trust, he was in a good position to do so. Interest in the old roses slowly grew, helped by nurseryman Peter Beales (1936–2013), who pioneered what became an international trend, starting a business selling old and species roses in 1968. David Austin (b. 1926) went one better and started to breed new roses that combined the beauties and scents of the old with the virtues of the new (repeat or continuous flowering, yellow shades, disease resistance); more than 190 varieties of these English roses, as they are called, have been introduced since Austin founded his business in 1969.

Such a popular flower inevitably carries huge levels of meaning. The revival of the old roses introduced a new level of social signification. By the 1980s, they had come (in the United Kingdom at least, with France and the United States following suit) to express the good taste of gardeners who wanted not only to conserve their garden heritage but to enjoy their rich fragrances and often subtle colours. Hybrid Teas and Floribundas came to be seen as scentless (many in fact were), garish, and above all vulgar—and no style-conscious gardener would be seen growing them.

Rosmarinus Lamiaceae

Rosmarinus is a genus of dry scrubland; the name is from the Latin for "dew of the sea." Its three species of subshrub are found in the lands circling the Mediterranean across to the Caucasus. The rosemarys are longer-lived than many Mediterranean region subshrubs, with some capacity for basal regeneration.

Rosemary has long been highly prized as both a culinary and medicinal herb. It is widely used in European Mediterranean cuisine for adding flavour, mainly to meat and poultry dishes but also to baked goods, notably the Italian focaccia. As an aromatic herb, it was much sought after during times of plague, when it was widely believed that the disease was spread through bad smells. Rosemary was woven into garlands for students sitting exams, as it was thought to aid memory—a use given some support by research: rosmarinic acid inhibits certain enzymes linked to neurological disorders that can contribute to memory loss. Rosemary also has strong antioxidant properties. Historically, it was widely used as a hair tonic, a mouthwash for teeth and gum health, a treatment for asthma, and much else. Rosemary oil, however, is very potent and needs careful prescribing, particularly during pregnancy. Rosemary ("for remembrance") was frequently used in funeral ceremonies across Europe and was customarily thrown into the grave in England during the 18th and early 19th centuries. Intriguingly, it also often made an appearance at weddings (a bride would wear it to symbolise her taking happy memories of her old home to her new one), baptisms, and first confirmations.

The strictly ornamental uses of rosemary have traditionally been somewhat secondary to its culinary and medicinal ones. Introduced by the Romans throughout their empire, it has been reintroduced many times since: as a borderline hardy plant, rosemary was frequently lost in the winter, and especially cold winters periodically eliminated it from northern Europe. Grown in France and Germany since the time of Charlemagne, it was largely used as a pot plant in monasteries until a wider distribution at the end of the Medieval period. Early gardeners often grew it on south-facing walls to minimise the risks of a hard freeze; one 16th-century source mentions it nailed to the walls of Hampton Court Palace, outside London. In 18th- and 19th-century Germany, it was an extremely common windowsill pot plant. Selection work during the 20th century has produced a range of plants of different shapes, prostrate ones being particularly sought-after, but there has been little fundamental change to the plant.

Rudbeckia Asteraceae

The name was bestowed by Linnaeus, in honor of his teacher at Uppsala University in Sweden, Olof Rudbeck the Younger, and his father, Olof Rudbeck the Elder. It is the dark centres to these golden-yellow daisies which draw us to them—hence, the common name black-eyed Susans (coneflower is another). There are 23 species, all rapidly growing herbaceous perennials and all native to North America, largely the midwestern and eastern U.S. states. Rudbeckias are generalists, growing in a wide range of both open and lightly shaded habitats, from virgin prairie to waste ground, but preferring moister and more fertile soils. Of the species in cultivation, *Rudbeckia hirta* and *R. triloba* are short-lived non-clonal perennials, often members of pioneer communities; the remainder are clonal and competitive, often strongly spreading, although not always persistent. Rudbeckias have had a minor role in Native American medicine, sometimes as a substitute for echinacea.

Rudbeckia laciniata was the first to arrive in Europe, to France in 1623. Another species, *R. fulgida* is of major commercial importance, thanks to its short stature (50–80cm) compared to that of most late-flowering yellow daisies; it shows considerable variation in the wild and has been divided into several varieties by botanists. The straight species came to England in 1760, and *R. fulgida* var. *speciosa* followed in around 1820; the latter quickly became popular as a border plant and cut flower. A selection of another variety, *R. fulgida* var. *sullivantii* 'Goldsturm', was one of the most frequently planted late-flowering perennials post-1980s; selected by one of his colleagues, Foerster named it and succeeded in bulking it up at his nursery just outside Berlin during World War II. *Rudbeckia hirta*, which also arrived in England in 1760, first received the attention of hybridisers in the 1950s, and by the end of that decade, Burpee Seeds had introduced the notable gloriosa daisies, with extra-large flowers. Hirta varieties have continued to be popular as annuals or short-lived perennials, with several seed strains now available in a range of yellows, orange-reds, and browns.

Salix <small>Salicaceae</small>

Salix (the name derives from the Latin for the tree) is a mostly northern hemisphere genus. China, eastern Russia, and the eastern part of central Asia are the centre of diversity, with nearly 400 of the approximately 450 species; a very few species are in South America and South Africa. Willows are thought to have originated in the Cretaceous. Nearly all willows have male and female flowers on separate plants, many of which are a short-lived decorative feature in the landscape. Species grown in ornamental horticulture are generally chosen for their foliage interest, or the intensely coloured bark of year-old stems, which can be a dramatic sight in winter. Three main subgenera are recognised—*Salix*, the true willows; *Caprisalix*, the osiers; and *Chamaetia*, the Arctic and mountain willows—but beneath these, taxonomic chaos reigns. Those in *Chamaetia* have wide appeal (although they are of particular interest to alpine gardeners), as their small size and attractive habit frequently makes them useful in confined situations. Willow identification and classification is hugely complicated by a very high tendency to hybridise.

Willows are popularly associated with wetlands; however, they have an ability to survive a wide range of difficult conditions, including low-nutrient soils and, paradoxically, drought. The narrow (and in some cases grey) leaves of willows are adaptations to drought, from which plants on riverbanks often suffer, if floods rip away large parts of their root systems or if waterlogging kills roots. It might be better to say that willows are found in wetlands because they can survive the problems associated with anaerobic root environments and physical damage better than most other woody plants. They are clearly stress-tolerators, but their tendency to distribute vast amounts of very fine seed shows their pioneer capabilities, while the ability of many (but not all) to regenerate from fallen trunks or broken-off branches illustrates competitive qualities. Their ability to rapidly build up large root systems is a useful adaptation to physically insecure environments. Many species are important for a

> It is a struggle; for though the black man fights passively, he nevertheless fights; and his passive resistance is more effective at present than active resistance could possibly be. He bears the fury of the storm as does the willow tree.
>
> (James Weldon Johnson 1871–1938)

variety of insect larvae, while the (predominantly wind-pollinated) flowers are also very good pollen sources for bees.

These are some of the most useful of all temperate zone plants for traditional societies. Wherever they are found, willows figure in traditional medical systems as a source of the mild painkiller salicylic acid (the precursor of aspirin), and the very resilient twigs of many different species have been woven into wicker, the raw material for baskets, hurdles, furniture, boats (such as the Welsh coracle), houses, cages, and a vast amount of other utilitarian items. Willows have traditionally been pollarded or coppiced in order to provide large quantities of the straight "wands" necessary for weaving; in many cases, these have become an important part of rural landscapes and, combined with the colour of the bark, an important garden feature. The twigs or bark have also been processed for papermaking or for tanning leather. As with many useful plants, willows appear in much folklore, although often with slightly sinister connections. They are associated with ghosts in Japanese rural lore, while in Europe, there were strong associations with magic, for example with Hecate, a pagan goddess with links to the underworld. Among European traditions one of the strangest is that willows were often supposed to get up and walk around by themselves at night.

The traditional use of willow as a living construction material has had a new lease of life through green engineering—for example, in making retaining walls and reinforcing riverbanks, so preventing erosion. Live willow can be cut and tied to form living sculptures or hedges—these enjoyed popularity in the late 20th century, but their lifespan tends to be short as the desire to grow into a tree soon takes over. This period also saw the rise of willow as a biomass crop for power generation.

As ornamentals, rather than sources of material, willow history is inevitably dominated by the weeping willow, whose origin is lost in the mists of time: *Salix babylonica* and similar taxa are thought to be Chinese; they were widely distributed through the Silk Road, finally making it to Europe, almost certainly via the Ottoman Empire, in the early 18th century. The poet and very innovative landscape gardener Alexander Pope (1688–1744) had a very early and notably famous specimen on his property west of London. Until the late 19th century, the weeping willow was the only one which would have been seen as ornamental, its familiarity given a strong boost by its appearing on willow pattern china, the most successful ceramic design of all time.

Exploration of eastern Asia brought a great many more species into cultivation, including some, such as *Salix fargesii* and *S. magnifica*, grown for their foliage interest. The late-19th- and early-20th-century rock garden craze brought many of the *Chamaetia* into cultivation, while a growing awareness of the need for low-maintenance winter interest in the latter part of the 20th century made gardeners look anew at species which in the past would have been seen as purely utilitarian, such as *S. alba*—a source of particularly good winter bark colour.

Salvia Lamiaceae

With a name derived from the Latin, *salvare* ("to heal"), it is clear that some of the sages have a significant medical history. All have a powerful aroma, very clearly that of Lamiaceae to any reasonably experienced gardener or botanist, but also very different from each other. Indeed, it would be fair to say that there is probably as much difference in aromatics from sage to sage as among the scents of any other genus. The range of colour is also unrivalled: intense true blues, deepest reds of every shade, black—only yellow is poorly developed. There are around 1,000 species, overwhelmingly herbaceous perennials or subshrubs, with a few biennials and annuals. Their main botanical characteristic is the presence of two stamens, which deposit pollen onto a pollinating insect through a lever mechanism during the younger growth stage of the flower; during the second stage, the female stigma becomes fertile and is able to pick up the pollen from the body of a pollinator. It is now thought that this mechanism evolved, at least twice, quite separately. DNA analysis suggests that the genus is composed of three separate

Salvia deserta in steppe grassland, probably former arable land, with *Origanum* sp. and *Linaria* sp., Kyrgyzstan. The species is very close to *S. nemorosa* and the other *Salvia* spp. of eastern Europe.

evolutionary groupings. No comprehensive study of the entire genus has ever been undertaken; any that will be will almost certainly result in extensive (and for gardeners, irritating) renaming. *Calosphace* is the largest of four currently accepted subgenera, it has 500 species, all New World, and all sharing a common origin.

Salvia is very widely distributed across the Americas, Eurasia, and Africa but absent from Australasia and the humid tropics. The Mexican highlands is a major centre of diversity (300 species). Other centres of diversity occur in the eastern Mediterranean region, western China, southern Africa (a smaller one), the northern Andes (centred on Colombia), and from southern Brazil across to Peru. A distinct group of around 20 species occur only in California. The Chinese species are particularly little known in cultivation. For the gardener, there are really only three useful categories (however little they may be supported by the botanists), and there are, needless to say, a good many oddities that do not fit in:

> [T]he exotic sages have no moderation in their hues. [T]he velvety violent blue of the one, and scarlet of the other, seem to have no gradation and no shade. There's no colour that gives me such an idea of violence—a sort of rough, angry scream—as that shade of blue.
>
> (John Ruskin 1819–1900)

- Herbaceous species of Eurasian origin. These are generally very cold- and drought-tolerant, e.g., *Salvia nemorosa*.
- Evergreen subshrubs of Mediterranean climates. These too are very drought-tolerant; included here are the familiar Old World *Salvia officinalis* and the very similar (and almost certainly distantly related) Californian *S. clevelandii*—an example of convergent evolution.
- Herbaceous and subshrubby species from Mexico and South America. These are borderline hardy and not notably drought-tolerant.

The Old World species in cultivation are generally plants of open habitats and which tolerate regular summer drought. Subshrubby species of the Mediterranean flourish in maquis and woodland edge habitats, while the herbaceous species are typical of dry meadow and steppe. The latter typically grow and seed within a short space of time but in moister climates can be persuaded to grow all summer and often flower twice. Most combine a stress-tolerant and pioneer character; they can colourfully dominate habitats in an early stage of succession. Eurasian species are noted as a good pollen source for bumblebees but poor for nectar.

Subshrubby American species are largely restricted to the coastal region of California and northern Mexico; several are characteristic of sage scrub, a variant of chaparral. Those species to be found in the mountain areas of Mexican southwards are more generally plants of woodland edge habitats and areas opened up by human activity. Many also flourish in mountainous cloud forest, and deciduous or mixed woodland at latitudes where light intensity at the forest floor is relatively high; their growth habit is stemmy and open, which gives a little insight into how they may perform in cultivation. American species are pollinated by both insects and hummingbirds; those with long tubular red flowers are most likely to be bird-pollinated. *Salvia coccinea* and others are an important nectar source for overwintering monarch butterflies in the Mexican highlands.

The complex aromas of *Salvia* species are created by volatile oils. These strongly deter predators, and in many cases appear to have an antifungal and antibacterial action, features which have no doubt been behind their popularity in a variety of herbal medical traditions. *Salvia sclarea* (clary) has a long history as an ingredient in perfumes and for flavouring wine. Of Mediterranean origin, *S. officinalis* (common sage) has

Salvia ×superba in Vaughan's 1936 catalogue.

(like rosemary) a long history of introduction and reintroduction to northern Europe. It appears in both Classical and Christian mythology, and in European culture is a plant of wisdom, health, success, and longevity. Widely used in herbal medicine, it is still used in remedies for coughs and colds, and in sage mouthwash for gum problems. In the past it was used for whitening teeth and appeared as an ingredient in early toothpastes. Brewers included it for its health-giving properties as well as for flavouring beer. It is one of the few species in the genus used as a common culinary herb.

Several species of the subgenus *Calosphace* are extensively used in herbal medicine and a range of spiritual practices in Native American traditions in Mexico, Peru, and Ecuador. One of them, *Salvia divinorum*, is used by Mazatec shamans in Oaxaca State, Mexico, and has proved popular with aficionados of hallucinogens. Unusually for a recreational drug, it appears to have a good safety record and low addictive potential and is hardly ever associated with crime. It is also generally legal.

Salvia hispanica (chia) is a Mexican annual species whose seed has long been used for food by indigenous communities, as an additive to maize flour; added to water, it makes a gelatinous mass, which can be used to prepare soup or drinks. It is highly nutritious and is showing potential for commercial cropping, particularly since its widespread adoption by the health foods industry in the 2000s. *Salvia apiana* of northern Mexico and the U.S. Southwest was also widely grown as a seed crop and medicinal herb; its foliage was made into bundles, which "smudge sticks" were burnt in purification rituals, a use revived by New Age spiritual practitioners.

Sage was written about by Classical authors as a medicinal herb and was probably first taken into cultivation in central and northern European monastic gardens, where there were no wild plants to harvest. The diversity of the genus was well recognised by the early 19th century, when British botanist George Bentham (1800–1884) listed 291 *Salvia* species in a monograph on the mint family (then called Labiatae); of these, probably only a fraction were in cultivation. During the 19th century, a steady flow of salvias entered cultivation in Europe and North America, but few became popular. The most successful was the scarlet *S. splendens*, discovered in the interior of Brazil between 1815 and 1817, and distributed across Europe soon afterwards. By the 1840s it was being used an ornamental, and it rapidly became one of the most widespread summer bedding plants of all time. Robinson listed some 40 species in 1909, but many of these were lost in the decades which followed. The 1970s saw the beginnings of a revival, particularly for the Latin American species, as gardeners in Britain and France began to appreciate the potential of plants that flower for a long time, in intense colours, and were just about hardy enough to grow outside with minimal protection. Gardeners in the warmer regions of the United States have also taken to the genus with enthusiasm.

Salvia nemorosa appears in *Hortus Eystettensis* (1614), but strangely there was minimal interest or selection of varieties of this and the other Eurasian meadow species until the postwar period. The modern history of this extremely resilient group of plants begins in 1945, with Foerster giving Ernst Pagels (1913–2007) a packet of seed. Pagels, who had just returned from a British prisoner of war camp and was trying to restart his career in horticulture, went on to breed many extremely good varieties from the 1950s onwards at his nursery in northern Germany.

Salvia splendens is still popular, particularly for the red component of patriotic planting schemes in countries with red, white, and blue flags. Once grown in vast quantities for Marxist-Leninist red-themed planting schemes, it has suffered from the demise of communism; democracy has been its downfall.

Saxifraga Saxifragaceae

Saxifraga is a varied genus of some 450 species, most best described as non-herbaceous perennials—nearly all have persistent above-ground growth but are in no way woody. The name comes from the Latin, *saxum* ("stone") and *frangere* ("to break"), as it was thought the plants could be used to treat kidney stones. Saxifrages are distributed across the higher latitudes and altitudes of the northern hemisphere, and then down the Andes to Tierra del Fuego; largely associated with mountain environments, they have become the core of the rock garden and alpine plant scene. Molecular genetic analysis suggests the family diversified during the mid to late Eocene, in either eastern Asia or western North America—indeed, the Bering Land Bridge would have meant plants could have dispersed either way. As the bridge disappeared during the Miocene, evolution continued separately. The genus has been divided into 15 sections, of which only five contain species regarded as horticulturally worthwhile or indeed realistically growable:

- *Irregulares* (formerly *Diptera*). With sprays of flower whose petals lack radial symmetry, and large leaves, these are distinctively lush-looking for saxifrages. Found from the Himalayas across to Japan and the Russian Far East.
- *Porphyrion* (including the former *Kabschia* and *Oppositifolia* sections). These usually form dense mats or cushions and have showy flowers. They have always been among the most important plants for the true alpine grower, being particular but not intractable, and therefore rewarding care with steadily growing tuffets that smother themselves with flower once a year.
- *Ligulatae*. Commonly known as encrusted saxifrages, owing to the white deposit of lime secreted by glands on the margins of the leaves. The rosettes eventually form dense mats or tufts. The flowers, nearly always white, are produced in conspicuous panicles. All are alpine, often growing on calcareous rock, quite often in shade.
- *Gymnospora*. *Saxifraga umbrosa* is a member of this small group of shade lovers, typically found in European foothills.
- *Saxifraga*. A large section of mostly cushion-forming perennials, including all the so-called mossy species that flourish in humid shade. Most are plants of shaded situations in alpine areas.

Saxifraga stolonifera was popular as a houseplant, its curtains of stolons with their "babies" very much part of its appeal. From *Gardening Illustrated*, 1881.

Saxifrages are superbly adapted, stress-tolerant plants, typically form-ing perennial rosettes of overwintering leaves, although some species are herbaceous and a few are monocarpic. Many form persistent clumps of highly integrated clonal individuals. Most are found in rocky mountain environments, although they vary considerably in their tolerance of sun exposure and drought. A number are species of arctic latitudes or are found only at very high altitudes, with no tolerance of heat. Others more typically occur in low-altitude shaded humid environments, where their clumps may be found on steep banks, wet rocks, and other situations that present phys-ical but not resource challenges.

Several species have had limited medicinal uses, and a few, most notably *Saxi-fraga micranthidifolia* (lettuce saxifrage) of the Appalachians, have been used as forage crops: "Both wilted saxifrage and saxifrage soup start a hungry man's nostrils quivering" (Angier 1974).

Saxifraga stolonifera was introduced in the mid-18th century from China. It became a popular houseplant, one which was hardy and easy enough for anyone, however poor, to grow; and with its mass of baby plants hanging on pink threads, it was simple, even fun to propagate and pass around the family or street. The related *S. fortunei* has long been popular as a garden plant in China and Japan, as suggested by its being named after Robert Fortune. Recent years have seen more breeding of *S. fortunei* in Japan and in Britain, exploiting this plant's reliably late-season flowering period.

Saxifraga 'Dentata' has many vernacular names, London pride becoming particularly common after a wartime song by Noel Coward. It is a cross between *S. umbrosa* and *S. spathularis*; natural hybrids are common in the Pyrenees where both are found, and the population in cultivation is made up of a num-ber of distinctly different hybrids. It was first mentioned by Gerard in 1597 and, since it coped well with polluted urban environments, went on to become one of those ubiqui-tous garden plants that is never sold, merely begged off a neighbour.

This huge race is the backbone of the rock-garden and no less.

(Farrer 1919)

The mossy species of section *Saxifraga* were once immensely popular, as they are easy compact plants and flourish in shade. A very large number of hybrids were raised in the 1890s, largely by Arends, later by a Mrs. Lloyd Edwards in Wales. The basic cross seems to have been *S. exarata* subsp. *moschata* with *S. rosacea*, although plants of this type had been in British cultivation since the late 1600s; others were added to the gene pool later. German and British breeders continued to produce many more until the 1930s. Most have now disappeared.

The alpine saxifrages, along with the rock garden, became popular in the 19th cen-tury. With the rise of the specialist alpine gardener some of these, notably the (former) Kabschia Group, were often grown into perfect hemispheres to be exhibited at shows, so smothered in flowers that the foliage was invisible. Hybrids of these and the related Engleria Group were raised and commercialised throughout the 20th century and despite a major decline in interest in alpines, several hybrids have been produced since the 1980s.

OPPOSITE Two popular rock garden sax-ifrages from the great era of the rockery, the red being *Saxifraga oppositifolia* and the white, *S. burseriana*. Illustration in T. W. Sanders's *Rock Gardens and Alpine Plants* (1910).

Sedums Crassulaceae

Sedum comprises around 600 species, united by their having succulent foliage; the name is from the Latin for "to sit," no doubt a reference to their habit. Sedums are found across the temperate and semi-arid regions of the northern hemisphere, with some in South America; the *Acre* clade in subgenus *Sedum* has a clear centre of diversity in Mexico. The taxonomy is uncertain, with relationships to *Echeveria*, *Aeonium*, *Sempervivum*, and other closely related genera still unresolved. *Rhodiola* has sometimes been included; this is a very similar succulent genus of high-altitude and high-latitude regions; *R. rosea* in particular has a very widespread circumpolar distribution. Most relevant to gardeners is the reclassification of the larger species as *Hylotelephium*, with 33 species distributed in Eurasia and North America.

All sedums are perennial, with some having a subshrubby habit. Many species have Crassulacean acid metabolism (CAM). The small-growing core *Sedum* species are evergreen rather than herbaceous, with a superficial root system, and have a tendency to creep, rooting from the stems; broken pieces too root easily, which may be a common method of propagation in nature. *Hylotelephium* and *Rhodiola* species are herbaceous perennials and form tight clumps with a considerable highly integrated root mass; hylotelephiums in particular are highly persistent and long-lived plants of dry meadows and scrubland. All sedums are stress-tolerators, found in a range of habitats that experience seasonal drought: dry grassland, rocky slopes, and open dry woodland. Core *Sedum* species tend to be restricted to very rocky environments where there is minimal competition from larger plants and where their ability to root from broken stems puts them at a clear advantage.

In English, sedums have a number of quaint country names of obscure origin, e.g., Jack of the butterie and welcome-home-husband-though-never-so-drunk. Many species have been eaten as salad crops in a variety of cultures, as they are succulent with a slightly acidic or astringent flavour (their name in German translates as "wallpepper"). Some however can be mildly toxic. *Sedum rupestre*, common in kitchen gardens during the 18th and 19th centuries in central Europe, was widely used as a salad green and in soups. *Rhodiola rosea* figures in herbal medicine, at least until the 18th century in Europe and more recently in China and Siberia. *Hylotelephium telephium* (orpine) often turned up in cottage gardens; it was said to carry on growing even when dug up and kept dry (hence another common name, livelong), so was sometimes used as an interior decoration. Some sedums, notably *S. alfredii*, are noted as hyperacumulators of lead, zinc, and cadmium, i.e., they absorb high levels of heavy metal pollutants and so can be used to help clean up contaminated soil.

In cultivation, the small species had some use in carpet bedding in the 19th century, and in grave plantings. Interest in them really took off only in the early 20th century, with the rise of the rock garden. The subsequent decline in interest in rockeries has done little to diminish that interest over the years, as the plants' small size and ability to grow in difficult habitats has ensured their continuing popularity, and many selections, usually based on foliage colour, have been made. Some native species, such as *Sedum morganianum* (burro tail), are popular as pot plants in Mexico.

Small, creeping sedums became enormously popular as a vital component of green roofing systems, which from a start in Germany in the 1980s have now burgeoned globally. Sedums have proved adaptable for both temperate and arid zone green roofs, as many species will survive in very shallow substrates; they have also become important in a range of modular ground cover systems, as they can be grown on mats that can be rolled up and transported like turf.

There was little interest in *Hylotelephium* species until 1955, when the Arends company released a hugely successful hybrid, 'Herbstfreude' ('Autumn Joy'), and the plants slowly gained in popularity. They have become very important as components of perennial plantings since; the extensive use of 'Herbstfreude' in work by the Oehme van Sweden practice in the eastern United States in particular has helped focus interest on the genus and stimulate interest in new selections. Currently, new selections and hybrids are produced at a constant rate.

Solidago Asteraceae

There are around 150 species of goldenrods, nearly all North American, with a few in South America down to the Southern Cone, and a handful in Eurasia. Goldenrods are often blamed for causing hay fever in humans; in fact the pollen causing these problems is that of wind-pollinated ragweeds (*Ambrosia* spp.), which flower at the same time. *Solidago* species do contain allergens, however, in the stem and leaf tissues, but this is likely to affect only those working professionally with the plants. All solidagos are clonal herbaceous perennials, often exhibiting features of both pioneering and competitive behaviour, but with varying rates of spread. Some of the more aggressively pioneering species have become important but rarely problematic aliens in continental Europe. In nature they are plants of woodland edge, marshland, and transitional grassland habitats. In mature prairie, however, they tend to be infrequent, unable to compete with the grasses. All are noted as exceptionally good nectar sources for honeybees and butterflies. The name is from the Latin for "to make whole," referring to their use in herbal medicine.

The goldenrod is yellow,
The corn is turning brown,
The trees in apple orchards
With fruit are bending down.

(Helen Hunt Jackson 1830–1885)

European *Solidago virgaurea* was renowned for a variety of medical uses (e.g., for healing wounds) and imported to England (no doubt at great expense) from the continent. But Gerard, in his *Herball* (1597), reported that after it was finally discovered growing in "Hampstead wood," it lost value considerably, and now? "No man will give halfe a crowne for an hundred weight of it . . . . This verifieth our English proverbe, Far fetcht and deare bought is best for Ladies." Native Americans also used goldenrods medicinally, although no one use stands out. The plants contain a rubber-like substance, which led Thomas Edison and Henry Ford to experiment with them as a commercial source of rubber, as did African-American plant scientist George Washington Carver (1864–1943). Ford did produce some tyres made of goldenrod-derived rubber, but in the end the project was abandoned, mainly because the rubber was of very low quality.

Tradescant the Younger had what was probably *Solidago canadensis* in his garden in 1634. It appeared in many other northern European gardens soon afterwards, spreading throughout Germany over the next century, alongside several other species. The botanic garden of Prague had 73 species in cultivation by the middle of the 19th century. During this time *S. canadensis*, or a hybrid of it, had become very popular as a cottage garden plant across Germany but does not appear to have started naturalising until the 1930s. A similar process happened in England, with railway embankments a particularly favourite habitat, often alongside *Symphyotrichum novi-belgii*—largely displaced by Japanese knotweed since the 1990s. Goldenrods were an essential part of the early-20th-century herbaceous border; a number of hybrids were raised and widely disseminated. By the end of the century, however, the plants had a bad reputation, in Britain at any rate; anything that gets to be seen on railway embankments or badly maintained pony paddocks will soon lose its popularity for the garden.

In the United States interest in growing goldenrods was at a peak in the 1920s, as part of a fashion for growing native plants, and during the 1970s interest in them grew again, as part of the revival of interest in natives and habitat restoration. Some selections are again being made; a good example of a new trend might be 'Wichita Mountains', a selection of an unknown *Solidago* species found in Oklahoma promoted for drought-tolerant planting.

Sorbus Rosaceae

Named after the Latin for *Sorbus aucuparia*, *Sorbus* currently holds 120 species of deciduous woody plants, but this is a problematic figure for a couple of reasons. One is that some botanists believe several species should be hived off into other genera, a move that, on the whole, is supported by evidence from molecular genetics. Another is that *Sorbus* includes many apomictic species, where flowers self-fertilise, with the result that any little genetic mutation can be propagated—the result is swarms of micro-species (80 in Europe alone). *Sorbus* is a part of the subfamily Maloideae of the Rosaceae, whose members (including *Malus*, *Pyrus*, *Crataegus*, *Pyracantha*, and *Cotoneaster*) are distinguished by having a distinct fruit "core"; even quite distantly related members of Maloideae seem to be able to cross. The non-pinnate species (i.e., the whitebeams), comprising 17 sexual and 30 apomictic species, are closely related to *Malus* and *Pyrus*, and the very distinctive *Sorbus torminalis* (service tree), to *Crataegus*.

The distribution of *Sorbus* is very similar to other genera with which it is often found (e.g., *Alnus*, *Populus*, *Picea*); there is a wide diversity in eastern Asia and a few but far more widely distributed species in northern Eurasia and North America. They possibly evolved in the Eocene in the Sino-Himalayan region (still the centre of diversity) and spread out from there. As trees of cooler climates they could only have spread out in cold periods; it is thought that the limited range of European and North American species are the result of post–ice age break-outs from the ancestral homeland.

Sorbus species vary from trees to alpine subshrubs, some suckering. They tend to be restricted to cooler, often montane, regions with higher rainfall. Most show a distinct pioneer habit and tend to be intolerant of competition, being largely eliminated in mature forest. An exception to this is the behaviour of the strongly suckering *S. torminalis*, found only as a component of ancient woodland. Plants have often been grown as grafts as the purity of the seed cannot be guaranteed; the downside of this is that the long-term performance of these species growing on their roots cannot be appreciated. Most perform better on their own roots, which in any case can sometimes sucker, so propagating the plant. The white- and pink-berried species are often apomictic and so can be raised reliably from seed. The brightly coloured fruit is very popular with birds, which generally eat orange-red fruit first, then yellow; only when all these are finished do they move on to white, pink, and crimson. Seed-eating birds such as finches also attack the fruit for seed but leave the flesh.

Sorbus aucuparia (rowan), the very widely distributed Eurasian species, was seen as a useful aid against witches and the devil in Medieval and early modern Europe. Not used as timber at all now, rowan once had a high reputation in Britain, as its timber was nearly all heartwood, and so good for tools and small high-value items. Its berries are unpalatable but can be made into a tart jelly, good for accompanying game; adventurous jam makers have used other *Sorbus* species to good, and colourful, effect. Since the 18th century there have been efforts in Russia and Moravia (now part of the Czech Republic) to develop more palatable varieties, with some success. Spirits distilled from the fruit have a long history in Austria, parts of Germany, and eastern Europe.

The Mountain-Ash, so esteemed among our northern neighbours as a protection against the evil designs of wizards and witches, is propagated by the Parisians for a very different purpose. It is used as one of the principal charms for enticing the French belles into the public gardens, where they are permitted to use all the spells and witcheries of which they are mistresses; and certainly this tree, ornamented by its brilliant scarlet fruit, has a most enchanting appearance when lighted up with lamps, in the months of August and September.

(Gilpin 1837)

Cultivation of *Sorbus* is a largely 20th-century development. Seedlings of *S. aucuparia* were picked out for garden and landscape use, with further selection of cultivars from wild collections of other species, mostly made in southern China, during the early decades of the 20th century. Selection has been made for fruit size and colour, quality of autumn foliage, and for fastigiate growth habit. There are some hybrids, including the familiar yellow-fruited 'Joseph Rock'. The latter part of the century saw many new introductions from southwestern China, but none have become important commercially as yet.

Spiraea Rosaceae

Once upon time this was a large genus, but the splitters have had their way, so *Spiraea* is down to around 80 woody plants found across cool temperate North America and Eurasia. Former affiliations, and therefore botanical closeness, are suggested by the common name meadowsweet, which Americans apply to *S. alba* and Britons use for *Filipendula*: all filipendulas used to be classified as *Spiraea*. Spireas are deciduous shrubs of woodland edge and open damp habitats. All are long-lived clonal competitive shrubs, and some are able to sucker strongly to form thickets. In Germany, large stands of spireas, or *Bienenweide* (literally a pasture for bees), were planted for the benefit of these pollinators from the 18th century onwards. Some species were used by Native Americans and early settlers as herbal cures, particularly *S. tomentosa* for dysentery. As with *Filipendula*, the plants contain salicylates and so have analgesic qualities. The genus is named after the Greek word for a plant used in making garlands.

Spiraea salicifolia is an east European species, introduced to Britain long enough ago (early 17th century) to acquire a few traditional English names (bridewort, queen's needlework). In Germany it became popular in the late 18th century in landscape parks and in hedging. The very similar *S. douglasii* from the Pacific Northwest (as far as Alaska) was also popular during this period. Today, both are hardly grown, although they can sometimes be found naturalised around old gardens; such plants are often a hybrid between these two, *S. ×pseudosalicifolia*.

A few other American species were introduced too, but it was the east Asian species, introduced during the 19th century, that made the biggest impact, combining a compact, non-running habit with freely produced flowers and a tendency to leaf colour variation; many cultivars were selected from *Spiraea japonica* in the 1800s with vast numbers being sold ever since, popular with both the landscape industry and home gardeners. *Spiraea nipponica*, *S. thunbergii*, and a hybrid involving the latter, *S. ×arguta*, positively smother themselves in white flowers and also rank as best-sellers.

Opposite *Spiraea douglasii* from the Pacific Northwest is named after the great explorer of the region, David Douglas (1799–1834). Illustration from van Houtte's *Flore des serres et des jardins de l'Europe* (1860).

SPIRÆA DOUGLASI *Hook.*

ƥ *Van Couver.* *Rustique*

II. ɪanv. 2.

Off. lith. & pict. in Horto Van Houtteano.

Stachys Lamiaceae

Named after the Greek for "spike," after the flower shape, this is a poorly understood and under-researched genus, with anywhere from 300 to 450 species of annuals, perennials, and shrubs found across temperate or montane Eurasia, into Australasia, Africa, and North America. There is little real clarity in the distinction between *Stachys* and other related genera; the familiar *S. officinalis* in particular does not appear to be closely related to the rest of the genus. Species in cultivation are notably long-lived clonal perennials, with varying rates of clonal spread. Some wild species are very effective at underground spread, enabling them to move and so adapt to changing environments. Habitats favoured tend towards the damp and open. Stachys are noted as good nectar sources.

> [T]he silver mats of *Stachys lanata*, so much nicer under its English names of Rabbits' Ears or Saviour's Flannel.
>
> (Sackville-West 1951)

As the herb betony, *Stachys officinalis* was widely used in traditional herbal medicine, at least in Europe, as were several species in Asia, most notably *S. affinis* (Chinese artichoke). Extracts have been shown to have antioxidant and antimicrobial activity, but there is no current use in conventional medicine. Several species, including *S. affinis* and the European *S. palustris* (marsh woundwort), have small edible tubers.

The woolly-leaved *Stachys byzantina* from southern Russia and the Caucasus, introduced (as *S. lanata*) in the early 19th century, took only a few decades to become a widespread garden plant. A small number of other species have become garden perennials. Most surprisingly, given its being a not-uncommon plant of western Europe, *S. officinalis* has only recently (c. 2000) become a garden plant of importance.

Syringa Oleaceae

Around 20 species are included in *Syringa*, a genus of woody plants, the lilacs, named for the Greek for "pipe," after the hollow stems. The species range across Eurasia, from eastern central Europe to Japan. All lilacs are found in regions with strongly continental climates and are notably very cold hardy. All are long-lived large shrubs or small trees, capable of basal regeneration. Most are found in woodland edge habitats or steep slopes, where tree growth does not impinge on light levels. Damp soils are favoured. The flowers can be accessed by many wild bees but are not popular with honeybees. Several North American moth species eat the foliage, even though the plant is not native.

Lilacs have entered popular consciousness largely because they are the first flowering shrubs of summer and, unlike rhododendrons, not so picky about soil. That they became much used in public parks and in the gardens of the rising middle classes from the 18th century onwards has given them a relatively high profile in both popular culture and fine art, reference to them often appearing in songs, novels, poems, and paintings (Russian painters seem to be particularly fond of them). Kate Chopin wrote a short story "Lilacs" (1894), while *White Lilacs* (2007) by Carolyn Meyer covers segregation in the South; *Lilacs* (2007) is a film about Rachmaninoff's life by director Pavel Lungin. In southern Sweden lilacs are used for hedging gardens around houses and even around fields; since several varieties are typically used, the effect can be very colourful when they flower.

Plants of *Syringa vulgaris* reached central Europe from the Balkans during the 16th century; *S. ×persica* had been grown in Persian gardens for centuries, as well as in the many countries influenced by Persian civilisation, arriving in Europe sometime in the 1600s. A few white and pink forms cropped up over the next few centuries. Lilacs were widespread by the end of the 18th century, mentioned as a hedge plant in northern Germany and in Sweden; they became a popular subject for song and poetry around this time, too. Loddiges of London introduced several new varieties in the 1820s, including a double white. By 1850 there were 25 varieties in cultivation.

Lilacs had an early success in both Russia and the United States (they were among the shrubs planted by Jefferson, non-native though they were). Early lilacs were all immensely popular for forcing, with large quantities being grown and traded across Europe for this purpose. By the end of the 19th century, the lilac was the most common garden shrub in Germany, alongside the rose, in the gardens of the poor as well as of the wealthy.

The late 19th century saw an explosion of interest, with 450 varieties recorded by 1928, and Lemoine the most important hybridiser. In 1843 Lemoine obtained a double-flowered sport from another grower; most of the lilacs he then bred were selections from seedlings grown from this plant or its descendants. Mme. Lemoine had better eyesight, so she would stand at the top of a ladder and carry out the actual cross-pollination. Lemoine's lilacs—among them 'Belle de Nancy', 'Mme. Abel Chatenay', 'Princesse Clementine'—somehow encapsulate France's Belle Epoque: decadent, flouncy, voluptuous. The company carried on breeding them, using different species, throughout the 20th century. By the time the firm shut down in 1955, they had introduced 214 varieties.

From *Syringa vulgaris* alone, nearly 2,000 cultivars have been derived. The early 20th century saw many more species arrive from China, often from areas with cold-winter climates such as Gansu and the Tibetan borderlands. They have helped make *Syringa* one of the most important genera of plants for use in severe-winter climate zones. Russian collections were also important, these being made from northern China and Japan. In Canada, Isabella Preston started work on them in 1920 at the Central Experimental Farm in Ottawa, largely bypassing the traditional gene pool based on *S. vulgaris*. Many of these now form part of the so-called Villosae Group (*S. villosa* was one of several species used), notable for having more delicately shaped flowers than the older gene pool and for blooming relatively late.

> Still grows the vivacious lilac a generation after the door and lintel and the sill are gone, unfolding its sweet-scented flowers each spring, to be plucked by the musing traveller; planted and tended once by children's hands, in front-yard plots,—now standing by wall-sides in retired pastures, and giving place to new-rising forests.
>
> (Henry David Thoreau, *Walden: or, Life in the Woods*, 1854)

Tagetes Asteraceae

Named by Linnaeus for an Etruscan god who sprang from a ploughed field, *Tagetes* includes some 40 species of annuals and short-lived perennials, the latter in some cases weakly woody, found in the warmer regions of the Americas. All *Tagetes* species are pioneer plants: the genus name reflects a link with the colonising of empty ground noticed by early botanical collectors. Naturally found in semi-arid habitats with plenty of gaps in vegetation for annual reseeding, they have benefited from human activity and are commonly found on waste ground throughout their range. Since their introduction there has been a long-standing confusion with *Calendula*, as both share the common name marigold. Strictly New World, *Tagetes* nevertheless includes the so-called African and French marigolds. Their pungent scent is thought to be largely defensive, helping the plants fend off many insect and animal pests. For the latter reasons, several species have been used as companion plants, particularly for Solanaceae crops like tomatoes and potatoes; one particular use in organic farming has been to help clear land of nematodes. *Tagetes minuta* has been used as a source of essential oil for the perfume industry and as a flavouring for food and tobacco. The flowers are popular food sources for pollinators of all kinds.

Tagetes species were regarded as the flower of the dead in pre-Hispanic Mexico, where they are still widely used in celebrations for the Day of the Dead. Their ritual use has spread to Hindu and Buddhist cultures too, replacing *Calendula*. Huge quantities are grown in India, Thailand, and Nepal for garlanding statues of the gods, shrines, and honoured guests. The psychoactive *T. lucida* has long played a role in the rituals of pre-Hispanic and Native American spiritual traditions; maximum potency is said to be achieved when it is consumed with peyote. Today, *Tagetes* species are being grown (mostly in China and India) for lutein production; this compound (a beta-carotenoid) is proving very useful for the treatment of macular degeneration, among other pharmaceutical applications.

Tagetes erecta was brought to Europe by the Spanish soon after the conquest of the Americas but was not initially widely grown there. It was taken up elsewhere, however, including North Africa; returning to Spain after a punitive expedition against the Barbary pirates (i.e., the Emirate of Algiers), it was dubbed the African marigold, a name which has stuck. *Tagetes patula* got landed with the equally inappropriate common name

OPPOSITE A single French marigold variety appears in an anonymously painted bouquet of summer flowers including two species each of *Centaurea* and *Viola*, *Convolvulus tricolor*, *Celosia argentea* var. *cristata*, and a double hollyhock. Dated 1680, origin unknown, assumed Dutch. Courtesy of the Rijksmuseum, Amsterdam.

of French marigold after being introduced to England by (French) Huguenot refugees in 1572. It appears to have been quite widespread by the 17th century across Europe. *Tagetes* species were grown as annuals and were reportedly very common in London gardens by the mid-18th century, but they were also disliked for their unpleasant smell. Intensive crossing between species began in the mid-19th century, largely to create plants for bedding-out with double flowers. Intensive breeding has continued since, largely for the production of highly predictably sized long-flowering plants for the bedding plant trade.

Marigold seed strains from Vaughan's 1936 catalogue, clockwise from upper left: Vaughan's Special Tall Mixed, Alldouble Orange, Little Giant, Guinea Gold. The last was recommended as "one of the best edging plants."

Trillium Melanthiaceae

Trilliums are well known for having all their parts in threes (the name is from the Greek for "three" and "lily"). Native to cool temperate regions of North America and Asia, the approximately 50 species of herbaceous perennials have suffered much reclassification over the years; once included in Liliaceae, they briefly had the glory of their own family, the Trilliaceae, before being included in Melanthiaceae. Members of the family, which include *Paris* and *Veratrum*, all seem to share very slow growth and longevity. Trilliums are the classic choice woodland plants, the kind of thing gardeners get very excited about if they are able to grow them successfully. In nature, plants may take 10 years to reach flowering size (in cultivation often only half this), displaying an extreme form of a stress-avoidance strategy, whereby an extensive root mass is built up over many years, which then gives the plant long-term resilience, as eventually large clumps may build up. Summer dormancy is a common reaction to drought. As with many woodlanders, the seeds are distributed by ants.

Being such a distinctive plant, and one which occurs in large numbers in suitable habitats in the eastern United States and the Pacific Northwest, the trillium has become something of an icon, with many businesses using the name "trillium" or the flower as a logo. Although the most frequent vernacular name is wakerobin, others (e.g., birthroot, birthwort) indicate a history of use in herbal medicine among both Native Americans and early settlers.

Trillium cultivation began in earnest in the early 1900s, when the growing of native American plants was fashionable as part of a patriotic celebration of the nation's natural beauties. Recent decades have seen extensive collection of seed from wild communities, which will probably play a vital role in the conservation of plants, which are very vulnerable to disturbance. Although little hybridising has been done, the selection of noteworthy variants and geographical forms is well advanced.

Trillium catesbaei in woodland, North Carolina Appalachians. The plant is named for English naturalist Mark Catesby, who played a key role in the European discovery of North America and the Caribbean.

Tropaeolum Tropaeolaceae

The name derives from the Greek for "trophy," a reference to the shield-like shape of the leaves (captured enemy shields were treated as trophies in Classical times). Members are commonly known as nasturtiums (not to be confused with *Nasturtium officinale*, a likewise peppery annual in another genus altogether). *Tropaeolum* species, approximately 80 annuals and perennials restricted to Central and South America, often have intense flower colours, not just scarlets and oranges but also blue and purples. Many species are long-lived clonal herbaceous climbers, often with tubers, some of them growing at high altitudes in the Andes, in woodland edge and scrub habitats. Species from the coast and lowlands are winter growers; those from high altitudes make growth in summer, and it is these alpine species that have a special following among connoisseur growers. The annual species tend to be plants of streamsides and damp ground; they were used as a peppery salad ingredient from very early on—the leaves contain high levels of vitamin C. The Andean *T. tuberosum* (mashua) is a perennial climbing plant grown from ancient times for its starchy tuberous roots.

Tropaeolum minus was introduced to Europe at the end of the 16th century, via Spain and its colony in Flanders; John Rea in the first edition of his *Flora* (1665) reported it as being "so well known that I need not to be curious in describing it, for few Gardens of note are without it." The plant in fact went through several ups and downs but was largely displaced by the more showy *T. majus*, which arrived in the late 17th century. Doubles of this species arose in Italy in the 1760s and were propagated as cuttings and widely distributed; a semi-double that came true from seed appeared in a California garden in the 1930s and was sold as the Gleam strain, with literally tonnes of seed of it being sold during this decade. The scarlet-flowering climbing perennial *T. speciosum*, introduced in 1847 by Messrs. Veitch, became in a few decades a member of the Scottish garden flora, as the climate was so similar to that of its temperate Chilean homeland. *Tropaeolum peregrinum* (canary creeper), a Peruvian native, was once very widely grown and continues to enjoy success as a half-hardy annual climber.

OPPOSITE *Tropaeolum peregrinum* in Vaughan's 1936 catalogue.

Tulipa Liliaceae

Approximately 80 wild tulip species are distributed from northwest Africa and the Iberian Peninsula across Eurasia to northwest China, with central Asia a clear centre of diversity. DNA analysis suggests they evolved during the Eocene. Horticulturally, tulips are classified using the following divisions, most of which are self-explanatory:

- Single early (Division 1).
- Double early (Division 2).
- Triumph (Division 3). Tall, mid to late season.
- Darwin hybrid (Division 4). Single large egg-shaped flowers on tall stems; however, older Darwin tulips are included in the next group.
- Single late (Division 5).
- Lily-flowered (Division 6). With pointed and reflexed petals.
- Fringed (Division 7).
- Viridiflora (Division 8). With a green midrib in the outer petals.
- Rembrandt (Division 9). With stripes and streaks, in imitation of the virus-infected tulips associated with the 17th-century Dutch tulip mania.
- Parrot (Division 10). With streaked and deeply fringed petals.
- Double late (Division 11).
- Kaufmanniana (Division 12). Derived from *Tulipa kaufmanniana*, short and early, the so-called waterlily tulip.
- Fosteriana or Emperor (Division 13). Derived from *Tulipa fosteriana*, early.
- Greigii (Division 14). Derived from *Tulipa greigii*, short and relatively large flowers.
- Species or Botanical (Division 15).
- Multiflowering (Division 16). With multiple blooms per bulb.

Tulips are overwhelmingly plants of extreme continental and often semi-arid climates. They are mostly spring ephemerals, growing rapidly in a short period defined by moisture availability. An additional factor is nutrient availability—the soils in which they grow are often fertile, but this fertility is only usable for the short period in which there is moisture. Competition for nutrients is often low, as the vegetation is so sparse; the implication for cultivation is that inadequate levels of nutrients often result in a failure to flower from year to year. Another reason plants in cultivation might fail to flower is that they do not get the summer baking that is necessary to ripen the bud tissue in the bulb. Tulips, unlike most garden bulbs, are consequently very often one-year-wonders in the garden. There is, however, considerable variation between varieties, and some will perform relatively reliably for several years. It is also worth noting that, unlike daffodils and many other bulbs, tulips are only weakly clonal, and do not form clumps of bulbs with anything like the same vigour. Some species, however, may seed extensively, resulting in them naturalising in suitable environments. This is most likely to happen in areas such as Colorado, which have a steppe habitat and climate similar to central Asia.

Tulips have traditionally featured in Persian and Turkish poetry, often as a token or symbol of love. They frequently appear in the visual arts of these cultures too, such as in miniature paintings and tiles. The narrow angular shape (reminiscent of the modern Division 6), however, is very different from the rounded, full stylization of a tulip that has always been more popular in the West; so the observer of Islamic art needs to "get their eye in" to spot it. The single flower, on top of its straight stem

OPPOSITE A 'Semper Augustus' tulip with two varieties of myrtle (*Myrtus communis*) and two shells, attributed to Maria Sibylla Merian, painted c. 1700. Dutch. Courtesy of the Rijksmuseum, Amsterdam.

gate maurine.

was seen, in the Ottoman world, to represent the letter *alif* (for "Allah") and therefore the unity and uniqueness of the monotheistic god. Beyond decoration, there is little herbal or other use for tulips, apart from being eaten, for example as a famine food by the Dutch in World War II. Today, the tulip has become very much a Dutch symbol—indeed, along with the windmill and wooden clogs, something of a cliché. The country is a major exporter of both bulbs and cut flowers; visiting the tulips fields in the Haarlem area is an important part of the Netherlands' tourism industry.

The history of tulips in cultivation is frequently complicated by confusion in the sources between tulips and other bulbs, but it is generally agreed to have begun in 10th-century Persia (Classical authors were mum on the subject). The plants were then taken up seriously by the Seljuk (whose ancestors originally came from central Asia) and the succeeding Ottoman Turkish empires, who adopted the tulip as an imperial emblem. Tulips were introduced to Europe by de Busbecq, the ambassador of the Holy Roman Empire to the Ottoman Empire, who in the 1550s sent bulbs to his friend, Clusius. Clusius took some plants to Leiden in the Netherlands in 1593, but they were stolen. Soon tulips were all over the Low Countries.

The extraordinary outburst of financial speculation in the province of Holland during the 1630s ("tulipomania") is well known. Although still theoretically under Spanish rule, the Dutch had been building up an extremely successful economy, largely through trade, with many of the key aspects of modern capitalism being invented in Amsterdam; however, given the country's geography, investment in land was difficult—so money had to seek other routes to grow. Tulips were one such speculative investment; they became status symbols for the newly rich merchant and financier class, which stimulated both a rise in prices and efforts to breed ever more exquisite blooms. Tulips were initially grown as florist flowers, classified as roses, bizarres, selfs, or bybloemens, depending on the distribution of ground colours and markings. The most sought-after bulbs were those infected with the tulip breaking potyvirus, which caused elaborate streaking in the petals. As far as the breeder was concerned, a tulip was only as good as its infection, which (since there was no understanding of either genetics or viruses) had to be left entirely to chance.

As the love of tulips grew, so did the prices, with particularly fine bulbs regularly reaching hundreds of florins, at a time when a pig might cost only 30. In 1636 a 'Semper Augustus' bulb sold for 6,000 florins. Given the inventiveness of financiers, who had made much of the wealth of Amsterdam possible, it is not surprising that a futures market developed—in bulbs not even planted. With tulips being traded on stock markets and the less wealthy members of society joining in, a speculative bubble formed. When traders could no longer get the prices they wanted for their bulbs, the bubble burst, and many were ruined. The opinions of historians and economists on tulipomania are mixed, some claiming coolly that the whole business was classic manic crowd behaviour, others that investors were behaving more or less rationally. For our purposes though, tulip fever marks a threshold—the first time in history that ornamental plants, and the diversity among them, become an important commodity, attracting the attention not only of connoisseurs but of society as a whole.

Other countries were not immune. English growers report that by the late 16th century it was possible to have three months of colour from a

> The beauties are always wearing tulips and roses
> Probably the flowers have lowered the prices of the pearls and jewels.
>
> (Sheikh-ul-Islam Yahya 1553–1644)

OPPOSITE 'Semper Augustus', the most highly prized of all tulips during the Dutch Golden Age, with roses, cyclamen, and an auricula, by Hans Bollongier (1600–1645), one of the earliest flower painters of the period, whose work helped promote both the tulip and city of Haarlem as a centre for the flower. Courtesy of the Rijksmuseum, Amsterdam.

good selection of tulip bulbs. A 1776 catalogue listed 655 varieties, and James Maddock (1822) makes a complaint that has been heard all too often down the years: "So fond are the Dutch of their money, that they forego all English improvements, rather than become purchasers of our new varieties, many of which possess as much merit as any of theirs." In Turkey, breeding had resulted in many fine interspecific hybrids well before their export to Europe, but during the 27-year reign (1703–1730) of Ahmed III, the Ottomans imported millions of the bulbs from the Dutch, stimulating a hybridizing frenzy. The sultan became obsessed with them and held extravagant tulip festivals, the costs of which contributed to fomenting a military rebellion, which ended in his abdication. Ottoman documents record some 1,300 varieties around 1730.

The key 19th-century change was the use of tulips as bedding plants for spring displays; the tulip went from being a pampered florist flower to an item of mass production. The innovation came from the Dutch firm of E. H. Krelage, who in 1885 bought up the stock of the Belgian nursery Lenglart, which was closing down. Krelage took the best tulips, and in 1886 launched them as Darwins, named for England's scientific genius. The Darwins are somewhat military in bearing, making them ideal for dotting colour into spring bedding. By the 1920s, London parks featured more than 200 British-grown Darwin types.

The other development in the 19th century was the opening up of the tulip heartland, the vast areas of steppe, semi-desert, and mountains in central Asia, then often referred to as Turkestan. Eduard Regel (1815–1892), director of the Imperial Botanic Garden in St. Petersburg, played a key role as intermediary between growers in western Europe and Russian collectors, who were fanning out in the wake of the Russian military, then occupied in brutally incorporating much of the region into the tsar's empire. By 1870, Dutch companies, most notably Van Tubergen, had seen the commercial potential and sent out their own collectors, who reached as far as Iran, Tibet, and the Altai mountains on the Siberian/Mongolian border. In 1897, the famously extravagant English plantswoman Ellen Willmott began to subsidise plant hunting expeditions into Persia, Armenia, and Turkestan. Large quantities of bulbs were dug up in the wild and exported to Europe; it has even been suggested that the apparent extinction of *Tulipa sprengeri* in Turkey was due to overcollecting.

By the early 1900s many of these species tulips were available in sufficient quantities to be used in the newly fashionable rockeries. Over the course of the century, cultivars with superior qualities for gardens were selected and, in some cases, have formed the basis for new classes of hybrids. An example of this process is *Tulipa fosteriana*, introduced by Van Tubergen collectors in 1904, which when crossed with *T. greigii* and *T. kaufmanniana*, produced a range of warm-coloured Fosteriana varieties. In some cases (and *T. fosteriana* is one), species have then been crossed with the pre-existing Darwin gene pool, to add traits for colour, vigour, and early flowering.

Verbascum Scrophulariaceae

Around 250 species, mostly herbaceous, are native to Europe and western Asia, with a centre of diversity in the western Mediterranean—70 species can be found in Greece and the Balkans. The name means "bearded" in Latin, a reference to the stamen filaments; tall varieties are commonly called mulleins. Most *Verbascum* species are biennial or short-lived pioneer perennials, occasionally annuals or subshrubs. *Verbascum thapsus*, a particularly successful pioneer plant, has spread far from its European home through accidental human agency, popping up in dry and waste ground locations all over the world; one plant can produce 250,000 seeds, which can survive buried for decades. Respiratory complaints seem to have been the focus for the use of *Verbascum* species in traditional European herbalism. The Romans dipped the seedheads into tallow and burnt them as a taper. The leaves produce a yellow or green dye.

Several species have been grown in gardens since the early modern period. Despite the overwhelming tendency of the genus to produce yellow flowers, other colours are known and it is from non-yellows such as *Verbascum chaixii* 'Album' (white) and *V. phoeniceum* (purple) that many garden plants have been bred, particularly the Cotswold Group, bred in England in the 1930s. Recently, corporate breeders have turned their attention to the plant—the short lifespan and some unusual colours make it attractive to the garden centre industry. The species are often seen in gravel gardens and other forms of naturalistic planting where spontaneous self-seeding is part of the management.

Verbena Verbenaceae

Verbena is a varied genus of some 250 species of annual and perennial herbaceous or semi-woody plants, the majority of which are native to the Americas with a few in Europe. The name derives from the Latin for the sacred boughs carried in procession by the Romans. The genetic history of the genus is complex: not only natural hybridisation but horizontal gene transfer marked its evolution; the latter—generally seen only in bacteria—involves the transmission of genes between individuals outside the normal reproductive process. Some of this gene transmission occurred with members of what is often now classified as a separate genus, *Glandularia*. *Verbena* species vary in their habitat preferences but are generally short-lived pioneer plants of open situations. *Verbena officinalis* prefers dry locations, whereas North American *V. hastata* is a wetland plant, as is the familiar *V. bonariensis*. The latter germinates on mud on riverbanks in South America at the beginning of the dry season but in suitable locations overwinters and can survive another year. All are good bee and butterfly nectar sources and can also attract hummingbirds.

Verbena officinalis (vervain) has a long history as a traditional and even magical herb. As a medicinal plant it was used for a very wide range of conditions, including mental conditions such as anxiety. Curiously several European languages associate it with iron (e.g., the German *Eisenkraut*). An essential oil from several species is used in herbalism and in the making of Verveine, a liqueur from Le Puy-en-Velay in France.

Verbena bonariensis was the first to make it to Europe, in the 1720s, but made little impact on gardens until the late 20th century, when it became popular as a self-seeding plant in naturalistic border plantings. More successful was the concentrated impact of several small scarlet species from South America (e.g., *V. peruviana*, introduced to England in the 1820s), and it was not long before it was "rare to see a garden or a balcony without them" (Loudon 1841). The resulting gene pool has always played a colourful but rather secondary role in bedding schemes. Recent decades have seen renewed investment in *Verbena* by corporate breeders such as Suntory and Syngenta.

Veronica Plantaginaceae (Scrophulariaceae)

Veronica, named after St. Veronica, a semi-apocryphal figure in 1st-century Palestine, includes around 350 species of annuals and perennials, mostly from the temperate northern hemisphere, mostly Eurasian. This is a very varied genus, with some species (e.g., *V. beccabunga*) happiest in wet mud, but with many being notably tolerant of dry soils. The centre of diversity (Turkey and the Balkans) has similarities with that of *Campanula*: plants are largely alpine and dry slope specialists, many of them cushion plants or very compact subshrubs. Most garden species are clonal herbaceous perennials, slowly clump-forming and with a distinct tendency to form massive, very fine fibrous root systems, which appear to monopolise the ground around the plant; *Veronicastrum virginicum*, a very long-lived and persistent prairie species until recently included in *Veronica*, shares this character. The non-alpine species are generally plants of grassland, steppe, and woodland edge, often dry. Flowers are blue or occasionally white.

Veronicas have been put to some minor medical herbal uses. They were recorded by early botanists but rarely cultivated until the late 19th century, and then largely in central Europe. From this time on, a number of selections from the species, usually based on variations in the intensity of the blue flowers, have been made for border and rock garden use. *Veronicastrum* appeared in Leiden Botanic Garden in 1720 and was reasonably widespread by the 1850s, although it was not until the 1990s that selections were made and it became popular, as part of the move towards naturalistic planting.

Viburnum Adoxaceae

Named after the Latin for the common European *Viburnum lantana*, this entirely shrubby genus has some 250 species: four from Europe and North Africa, 20 from North America, 60 from Mexico southwards along the lower slopes of the Andes, and the remainder from Asia, in mountain regions as far south as Papua New Guinea; there are none in Africa. The centre of diversity is southern China. The genus has been divided into several sections, but many species remain unclassified. The genus appears to have evolved in the Paleocene, in the Old World, spreading later to the Americas. Viburnums in the wild are plants of scrub, light woodland, and woodland edge habitats. Being evergreen indicates either an origin in a Mediterranean climate (e.g., *Viburnum tinus*), where the growing season is limited to winter by summer drought and heat, or woodland understorey (e.g., the Chinese *V. davidii*). All species are truly long-lived woody perennials, with considerable capacity to regenerate from the base; some may sucker extensively, too. Viburnums are also noted as a high-quality nectar source for bees, but they can be poor for pollen. Winter-flowering ones are clearly good for early pollinators. Uses have been minor; *V. lentago* berries can make an acceptable jelly, while *V. dentatum* has been one of several species used for making arrow shafts.

The sterile form of *Viburnum opulus* (snowball tree) was known in cultivation by the

A flowering stem of *Viburnum lantanoides*, growing in the North Carolina Appalachians. The very stemmy growth habit of the plant is typical of many shrub species growing as woodland understorey. In garden conditions there is usually more light, and a more compact growth habit results.

mid-16th century; hailing from the Low Countries, it brought a Dutch name with it, which was anglicised as Guelder rose (and then attached to the entire species). Heads of sterile florets made it unique at the time, and so it was widely distributed, being used on a very large scale during the era of the landscape park in the 18th century. More highly valued, however, in the early modern period in England was *V. tinus* (laurustinus). It was hardy enough to be grown outside (often as a hedge), was easily clipped, and flowered early in the year. In colder climes, it was kept in pots and brought inside with the myrtles, oranges, and other precious evergreens.

These two species were pretty much the only ones grown until American natives began to appear in Europe in the 18th century, but they missed their warm summers and did not perform well. As with many other genera, it was the east Asian taxa that got gardeners excited: *Viburnum plicatum* f. *tomentosum* 'Mariesii' from Japan, *V. carlesii*, discovered by the British vice-consul in Korea in the 1880s (and named after him), and *V. farreri*, a popular garden plant in China; these three offered fine foliage, scent, and winter flowering, respectively. Some basic hybridisation often had to follow to iron out the less satisfactory aspects of the natural species and turn them into good garden plants: *V.* ×*carlcephalum* and *V.* ×*bodnantense* are far more familiar to us than *V. carlesii* or *V. farreri*, for example. But Wilson's introduction of low-growing evergreen *V. davidii* in 1904 has brought a good-looking plant, which needed no improvement, to many a garden and supermarket car park.

Much recent breeding has been done in the United States, where good new selections and crosses have been made from the Asian gene pool, some new species have been introduced, and the native species of North America themselves further sifted for garden-worthiness.

> Have you got *Viburnum carlcephalum*? If not, please get it at once. . . . It is one of the most exciting things I have grown for years past.
>
> (Sackville-West 1951)

OPPOSITE Heads of *Viburnum opulus* 'Flore Pleno' are the centrepiece of an opulent still life by Georgius Jacobus Johannes van Os (1782–1861), along with carnations, violas, lilies, *Scabiosa atropurpurea*, what could be freesias (top), and possibly some peonies. Oil on wood panel. Courtesy of the Rijksmuseum, Amsterdam.

Vinca Apocynaceae

The genus, whose name derives from the Latin for "to bind," used to encompass the largely African and Indian *Catharanthus*. The five species that remain in *Vinca* are native to Europe, northwest Africa, and parts of western Asia; all are well adapted to a life in woodland and woodland edge habitats around the Mediterranean. They are technically shrubs rather than herbaceous, as their above-ground growth persists from year to year, and they are true evergreens, with leaves living for more than one year. Their habit of spreading by stolons enables them to run over the top of other vegetation and then root when the stolon hits the ground; this gives them a competitive advantage, and not surprisingly they are regarded as invasive aliens in the United States, among other places.

Periwinkles, as they are commonly known, are often indicators of former human habitation, as the poem demonstrates (see sidebar). The Romans used the plant to make wreaths, a practice which continued in some traditional southern European cultures until recent times. Probably in cultivation in central Europe, and possibly England, since the 14th century, they were grown in herb gardens as medicinals: at least 80 alkaloids can be extracted from *Vinca* species, and the closely related *Catharanthus roseus* (formerly *V. roseus*) has produced a drug that is extremely useful in cancer chemotherapy. Their evergreen nature and willingness to grow in shade made them popular for growing on graves, particularly the smaller *V. minor*, which is still popular in German cemeteries. This also appears to have been a popular garden plant from very early on in the early modern period among town burghers and richer peasants. A double appeared in Flanders in 1576, and from the 18th century on, numerous colour and variegated forms of both *V. minor* and *V. major* were spotted and grown on as cultivars.

> There once the walls
> Of the ruined cottage stood.
> The periwinkle crawls
> With flowers in its hair into the wood.
> In flowerless hours
> Never will the bank fail,
> With everlasting flowers
> On fragments of blue plates,
>   to tell the tale.
>
> (Edward Thomas 1878–1917)

Viola Violaceae

Botanists reckon there to be around 525 *Viola* species. The name derives from the Latin for a number of scented flowers. Old enough to get itself distributed globally, including Australasia and many tropical mountain ranges, *Viola* is thought to have originated in Andean South America during the Oligocene, possibly earlier. Although most species are northern hemisphere, around a quarter are South American, including some extraordinary plants to be found in cold high-altitude deserts: the rosulate violas, which form very tight rosettes or tufts high in the southern Andes. *Viola lilliputana*, a rarely seen species of the Peruvian *puna*, survives vicious winds, intense cold, and solar radiation; the entire above-ground portion of the plant is just 1cm high. *Viola* species are overwhelmingly herbaceous, although Hawaii has some shrubby species, and there is the woody Iberian *V. arborescens*. Plants range from long-lived clump-forming perennials to annuals and other short-lived pioneer species. As with *Primula*, the ability to grow at low temperatures is strong: plants can thrive during winter in regions where temperatures

Violas and pansies do not appear much in early illustrative material, perhaps reflecting a fairly lowly status in the garden world—as cheap, hardy, and easy plants, they were very much plants of the urban poor. This is a painting on what appears to be alabaster, dated 1893, probably by an amateur artist, Eva Nathan, presumed British.

allow some growth and go on to flourish and often complete their lifecycle in spring. A great many are woodlanders, often preferring damp situations. Many are found in prairie and steppe, where they are most likely to be summer dormant. This early growth period means that many species can be grown in close proximity to later-flowering plants, either as perennials or as continuously sowing annuals.

Various legendary origins have been proposed for violas, including their springing up from wherever the mythical Greek musician Orpheus put down his lyre. In folk culture, violas traditionally symbolise humility, death, rebirth, spring, and love. Surprisingly, then, their reputation has tended to be rather sinister in Christian Europe; for example, they were used to garland criminals on their way to execution in Medieval times. In Europe, violets have long been used as a flavouring and colouring agent for food, with flowers for such purposes being dried or preserved in syrup or oil. Violet butter was promoted for students by Culpeper (1652) as it "rejoiceth the heart, comforteth the brain, and qualifieth the heat of the liver." Crystallised violets and violet-flavoured fondant fillings for chocolates are the most familiar culinary use today, but historically violet flowers have been used in salads, soups, ice creams, cakes, and jams, and in Italy the leaves lightly cooked as greens. Blue syrup of violets was used extensively by early chemists for testing pH, as it turns green in alkali solutions and red in acid—violets were once grown on a large scale in the English county of Warwickshire for this purpose. Plant material has been used to treat a very wide range of medical and psychological complaints, but with little evidence of any genuine effect.

Viola species have been widely grown in gardens, probably first by the Romans. A tendency to naturalise from old gardens has meant that populations have flourished well north of their wild range.

OPPOSITE "Mammoth butterfly pansies" offered in the Peter Henderson and Co. catalogue of 1897. Basic pansy forms and colours have changed little since, but big strides have been made in flowering period, reliability, and compactness.

Sweet violets

The species with the longest history in cultivation is *Viola odorata* (sweet violet), known for its delightful but evanescent perfume—smell it a second time, and it has gone! Ionine, the compound which is the heart of the scent, has an anaesthetic effect on the organs of smell. Grown by the ancient Greeks, it was a symbol of the city of Athens. It was cultivated by the Arabs, the Ottomans, and almost certainly in the European Medieval period from very early on; Hildegard von Bingen mentions it. Doubles and varieties in different colours are known from the 16th century. Cultivation started in earnest around 1700 near Paris, as well as in the region around the perfumery centre of Grasse in Provence. Napoleon loved the flower, picking some from Josephine's grave to take them with him to exile in St. Helena.

Parma violets, with the finest scent of all, are not hardy and need careful cultivation. They are of rather mysterious origin (albeit with a historic connection to Naples); it is possible they originated in the Middle East as a hybrid, as detailed instructions on growing similar-sounding violets are known from 10th-century Arabic texts.

Scented violets were extremely popular in 19th-century England and France, with large quantities being grown in favoured localities, or under glass, for transport on the railways to cities, where they were popular as boutonnieres and in small arrangements (e.g., for table decoration). In North America, they were grown only in the vicinity of a few large population centres; the town of Rhinebeck in the Hudson Valley once had over a hundred businesses (mostly run on a part-time basis) growing the flowers for sale down river in New York City. The turn of the century was the sweet violet's high point, with a steep decline after the 1930s, almost to insignificance. The French city of

PRICES
OF SEEDS:
Any Variety shown
on this Plate
15¢ per Pkt.
Buyer's Selection of
any Six Varieties 75¢
or the entire
JUBILEE COLLECTION
OF
MAMMOTH
BUTTERFLY PANSIES
(12 Varieties) for $1.25
All Colors Mixed 30¢ per Pkt

PETER HENDERSON & CO'S
"JUBILEE" COLLECTION
OF
MAMMOTH
BUTTERFLY PANSIES

COPYRIGHT 1897 PETER HENDERSON & CO. N.Y.

SCHUMACHER & ETTLINGER. N.Y.

Toulouse was a major centre, with the violet trade continuing well into the 20th century. Today a range of violet-based food and other products are sold there; the plant is also the city's symbol. In some cases, as around the towns of Clevedon and Dawlish in south-west England, escapees from cultivation can still be found in the hedgerows.

Pansies

The origin of the word "pansy" is the French *pensée*—a thought, in the sense of a memory and remembrance. Perhaps unusually, the pansy (*Viola ×wittrockiana*) has a very clear historical origin, with two aristocratic garden owners and their head gardeners developing the plant more or less simultaneously within a few miles of each other, to the west of London, although we know nothing about what, if anything, passed between them.

Viola tricolor, an annual of arable land and well known to early gardeners, is one of those plants that attract a vast number of vernacular names, many rather colourful. Heartsease is the most common; many others have an amorous or sexual bent: tickle-my-fancy, Johnny-jump-up, Jack-jump-up-and-kiss-me, come-and-cuddle-me. In the early years of the 19th century Lady Mary Elizabeth Bennet began to collect all the *V. tricolor* plants she could get hold of and set them out in her father's garden at Walton-up-on-Thames, Surrey. She was probably drawn to the species by the high level of genetic variation it shows; other species (e.g., *V. lutea*) were also possibly involved in what would become *V. ×wittrockiana*. With the help of her head gardener, she carried out cross-breeding and produced a range of distinct varieties, in which a number of nurseries took an interest from c. 1812 onwards. Oddly enough, Admiral Lord James Gambier was doing the same with his gardener, just down the road; the latter gentleman, a Mr. Thompson, "father of the pansy," reportedly produced vast numbers of varieties, which he too passed on to nurseries.

Public popularity followed swiftly, helped by the fact that a well-known actress, Sarah Siddons (1755–1831), fell for the flower and filled her garden with them. Darwin noted 400 named cultivars in 1835. New introductions, such as the Russian *Viola altaica*, brought in fresh genes, while occasional mutation provided new characteristics, such as when a self-sown seedling of Thompson's appeared with distinct blotching—this became the ancestor of the modern pansy. Interest was reflected in some very high prices and a desire to show the plants competitively. A Hammersmith Heart's-ease Society was founded in 1841 and the Scottish Pansy Society in 1845. Show rules were very rigid, however, demanding almost circular flowers, a clearly defined border on the lower petals, etc. Another range of plants had been bred in France and Belgium to be used as garden plants rather than show pansies; they were more varied in form and colour, with the large velvety blotch we are familiar with. John Salter, an enthusiastic cultivator of florist flowers, introduced some of these in 1847, and after a slow start, these Belgian or "fancy" pansies gained popularity, largely displacing the show pansy by the end of the century. These were also crossed with *V. odorata* to produce varieties suitable for bedding out.

In the 1860s, Scottish nurseryman James Grieve (of the famous apple) crossed show pansies with *Viola cornuta* and possibly other species to produce a range of compact, tufted, and longer-lived plants—these came to be known as the violas. A decade later, another Scot, a Dr. Stuart, bred a range of violas that lacked the rays on the petals; these he called violettas. *Viola cornuta*, a relatively reliably perennial species, has since become of considerable garden importance on its own merits.

Pansy (and to a lesser extent, viola) production is now a major commercial business, very largely for the bedding plant trade, as the plants offer one of the cheapest and hardiest possibilities for winter or early spring flower. Breeding continues to aim for bold flowers, a long flowering season, weather-resistance, and consistency.

Vitis Vitaceae

Vitis (from the Greek for the plant) includes approximately 60 species. The distribution is almost entirely temperate northern hemisphere with some penetration into South America. It is thought the genus diversified in the Eocene, with some species evolving into geologically recent times as climatic and physical barriers separated them from each other. There is a preponderance of Asian species with a centre of diversity in south-central China. Nearly all *Vitis* species are lianas, are wind-pollinated, and have roots that can store considerable quantities of nutrients during the dormant season. They form an integral part of forests, climbing up trees to produce their leaves, flowers, and fruit often tens of metres above ground level. They are particularly lush in woodland edge habitats, where their ability to cover trees may result in them being visually dominant.

Vitis vinifera, a species of the Caucasus, is one of the most important of all fruit crops, a source not only of fresh grapes but also of wine, raisins, and currants. It interbreeds easily with other species, and historically many North American grape varieties are hybrids between it and other species, such as *V. labrusca*, *V. aestivalis*, and *V. riparia*. The use of non-vinifera grape varieties has long been a matter of conflict within the wine industry, with French growers traditionally bitterly opposed to them. *Vitis coignetiae* grapes are used to produce wine in Japan and Korea. The stems have also been widely used in basketmaking by traditional cultures. *Vitis coignetiae* has a particular place in Japanese culture owing to its popularity with artists, for whom it has long been a favourite subject for ink painting.

The appreciation of the grape vine simply as an ornamental plant is a recent interest—a few centuries, compared to the thousands of years the plant has been cultivated for its fruit and its fermented juice. A few purple- or cut-leaved forms were grown from the 17th century onwards, some American species were introduced to Europe (and largely ignored), but in the mid-19th century the spectacular species of eastern Asia became known. They proved difficult to propagate until seed arrived, however, so the best known, *Vitis coignetiae*, did not take off commercially until the very end of the century. Some other species have been introduced more recently.

明治廿六年四月五日印刷
年四月八日出版

版權所有

大阪市西區立賣堀北通四丁目三番屋敷
發行兼印刷者　青木恒三郎

Vitis sp. in autumn colour; *Vitis coignetiae* was a popular subject for Japanese artists. Print by Kono Bairei and Aoki Kosaburo, of Osaka, dated April 1893. Courtesy of the Rijksmuseum, Amsterdam.

Wisteria Fabaceae

Wisteria comprises 10 species of woody climbers native to the eastern United States and eastern Asia. The spelling has been disputed, and "wistaria" is sometimes seen in older sources: the name honors Caspar Wistar, an 18th-century U.S. physician who had nothing to do with the plant and very little to do with botany. Wisterias are long-lived lianas, able to reach the top of canopy trees, and therefore very different from the smaller climbing plants more familiar in gardens. In the United States, both the Asian species in cultivation, *Wisteria floribunda* and *W. sinensis*, have been noted as potentially invasive.

Wisteria stems have been used for rope and basket making, and in China even as a source of fibre for weaving fabric for making clothes, usually rough working ones but also for mourning, as the plant is associated with longevity. The pendant flowers are frequently featured in Far Eastern art; after their introduction to Europe, plants became very popular subjects with Western artists as well.

Plants were long cultivated in China. The building of trellises to support the plants instead of letting them ramble over trees or buildings seems to have started in the late 17th century. During the Heian period in Japan, wisteria viewing parties were popular, especially since the flowers were a symbol of the ruling Fujiwara clan. There was some selection during the Edo, focussed chiefly on the length of the raceme. A plant at Kyushaku in Saitama Prefecture is considered the largest and oldest wisteria in Japan; it fills a 990m^2 trellis and is thought to be 600 years old. The largest plant in North America, planted in 1894 at Sierra Madre in California, has been the centre of an annual festival since 1918.

Wisteria floribunda was spotted by Fortune in the garden of a Chinese official in Zhoushan. Fortune was able to obtain plants in a local nursery and sent them home in 1845; the *Gardeners' Chronicle* noted that the new species had even attracted the attention of Queen Victoria herself. Others soon followed, with much hybridisation being carried out in France. Breeding has been relatively low-key since, with the focus on raceme length and colour range. Breeding and selection has made almost no difference to the basic form or size of the plant.

Yucca Asparagaceae (Liliaceae)

Potentially hardy members of this 50-species New World genus can be found as far north as Alberta in Canada, as far east as Virginia, and as far south as Guatemala. Non-hardy species can be found in arid areas of the Caribbean and in South America. The name is the result of a confusion created by Gerard, who obtained a plant in 1593, calling it yucca, after a Caribbean name for cassava, with which he (or his source) confused it. Yuccas are found in a wide range of habitats, generally arid: scrub, semi-desert, steppe and other dry grassland, coastal sands, and in high-altitude arid environments where nighttime winter temperatures can be very low. The plants' growth habit may look very different from that of broadleaved shrubs but is essentially analogous to that of clonal shrubs, usually with only very slow regeneration of the plant from the base. They have a very specialised pollination system, based on a mutualism with yucca moths (family Prodoxidae), which not only transfer pollen but also lay eggs in the flowers, the larvae then feeding on some of the developing seeds, a price the plant is clearly prepared to pay.

Native Americans have very extensively and imaginatively used the tough fibres of the leaves for twine and rope, sandals, textiles, and basketware (for which the Navajo are notable today). The roots contain saponins, which means they can be used directly as a soap, or for making soap and other products; modern industrial applications include the manufacture of foaming agents in carbonated drinks. The seeds, unripe or ripe, have long been consumed, either ground into meal or baked; they are still widely eaten in Central America. The core of the plant can be cut out and cooked, as the agave is used to make mezcal and tequila. The flowers too are edible, and the fruit has been fermented and used ritually by some Native American groups. Various plant parts have been used to treat a wide range of symptoms in traditional Native American medicine systems. There is some evidence that yucca extracts may help in treating arthritis and hypertension.

The yucca was northern Europe's first real encounter with reliably hardy exotica, *Yucca gloriosa* arriving in England in the 1550s. *Yucca filamentosa* came in 1675, followed by a trickle of other species ever since. Various hybrids have been bred over the years, in Britain and in Germany, but very few have survived. Doubtless several, more cold hardy species could be grown outside their homeland. Slow propagation has been a possible obstacle to the wide dissemination of new species.

William Robinson was a great exponent of dramatic and exotic foliage. Here, from the 1921 edition of *The English Flower Garden*, is a scene from Abbotsbury Gardens, in a sheltered location in the southern English county of Dorset: a large planting of a *Yucca* sp., with what looks like *Cordyline australis* and *Aralia elata*.

Zinnia Asteraceae

Named after 18th-century German botanist Johann Gottfried Zinn, the genus includes 17 species, all found from the southern United States into South America, with the most diversity in Mexico. *Zinnia* species are found in open, often dry habitats. Their rapid growth and short lifespan (annual or short-lived non-clonal subshrubs) illustrate a stress-avoiding nature in harsh environments. *Zinnia grandiflora* is used in herbal medicine by several Native American communities in Mexico.

The plants were grown by the Aztecs, and introduced into Europe by the Spanish in the late 18th century. A range of colours was known from early on, probably a legacy of Aztec breeding. Ancestral species vary in colour—*Zinnia elegans* is purple, *Z. peruviana* red-orange, *Z. angustifolia* and *Z. grandiflora* yellow. One of the strong points of the plants as ornamentals is the ability to get vivid purple, pink, yellow, and orange, all from the same seed packet. They were very popular in the 19th century, suffered something of a decline in the late 20th, and now seem to be on the way up once again. Some of the most adventurous breeding is being done in Japan, for the cut flower trade.

Epilogue: These We Have Lost

In writing this book I have been made very aware of plants we no longer have in our gardens. Any look through historical material makes one realise just how much we have lost, as well as gained. Here, I would like to flag up some of the most important losses, partly just to encourage people to find out more about the plants but also to encourage gardeners to try some of them again.

Consider the array of plants available to gardeners in Britain during the latter part of the 19th century, and to a slightly lesser extent in France and Belgium (and distinctly lesser in Germany and the United States): one cannot help but notice the massive reduction in available warm-climate species since that time. Anything that needed protection under glass and heating was likely to become a victim of rising fuel and labour costs, to say nothing of changes in fashion. The result is that we no longer grow, or even know, vast swathes of tropical flora. More surprising, though, has been the loss of species that need only frost protection. I am thinking here particularly of the flora of the South African Cape, which was the first overseas botanical wonderland to be discovered by Europeans, from the 17th century onwards. A great many Cape bulbs were introduced and then lost, but most spectacular has been the rise and fall of the Cape heaths, those members of the genus *Erica* whose evolutionary explosion in the Cape Province is perhaps the most extraordinary example of the capacity of plants to diversify and speciate.

The development of the first greenhouses in the late 18th century created the opportunity to grow plants from Mediterranean climates. Simultaneously, the British (or at least the tiny minority who could afford a glasshouse) developed a craze for Cape heaths. A very wide range of species were embraced, along with a number of hybrids—their exotic colour combinations (many are deep yellow or a mix of red and yellow) and the tendency of many to flower in the winter must have made them irresistible. By the 1820s they were being displaced, as warmer climate zones were being explored and greenhouse technology was now so much better that plants from these climates could now be safely grown.

Epacris longiflora featured (as *E. grandiflora*) in the 1806 edition of *Curtis's Botanical Magazine*. This magazine, which still runs, was, during the 19th century, the main source for images of the endless procession of botanical riches which European exploration of the globe was revealing.

The number of species declined steadily, and by World War I Cape heaths were very much a minority interest; a few were grown as Christmas pot plants until the 1970s (I remember, as a teenager, keeping two going for several years). A revival started towards the end of the 20th century, in which I probably played a modest role, as my very first article in a garden magazine was about Cape heaths; gardeners in sheltered places in Cornwall and Brittany have since discovered that they can grow them outside.

In Australia, *Erica* is replaced by the very similar and related *Epacris*, species of which became widely grown as cool greenhouse or conservatory plants in Europe during the 19th century, with around twenty cultivars selected. Their disappearance has been total: not one of the cultivars survives, the plant is hardly cultivated in its native Australia—and in fact, I don't think I have ever seen one, not even in a botanic garden.

Another major Victorian favourite that has vanished as an ornamental plant but which survives in the floral industry is *Bouvardia*, a Central American genus of loose scrambling plants, the white species of which are heavily scented (to attract pollinating moths). Victorian conservatories tended to be full of cast-iron pillars to support the roof, so climbing or scrambling plants like bouvardias filled a "cultivation niche"—for a

OPPOSITE From the 1924 edition of *Kelway's Manual*, photographs of four *Gaillardia* and *Tanacetum* (then called *Pyrethrum*) cultivars. Courtesy of Kelways Plants Ltd.

"Few plants are more easily grown, or more amply reward the care of the cultivator . . . than the bouvardias," says *Gardening Illustrated* in 1881. This bouquet includes some of the many species and cultivars that were grown at the time.

PYRETHRUM, LORD ROSEBERY (Kelway).
2/- each.

GAILLARDIA, KELWAY'S KING.
2/- each.

GAILLARDIA, SUMMER (Kelway).
1/6 each.

PYRETHRUM, LANGPORT SCARLET (Kelway).
5/- each.

KELWAY'S NEW PYRETHRUMS AND GAILLARDIAS.

time. Many cultivars were selected. Not a single one survives in general cultivation.

A great many annuals or short-lived plants in cultivation in Victorian times are still with us but greatly reduced in number. In particular, those annuals which are not hardy and were grown as late winter-flowering pot plants have seen major declines. *Pericallis* ×*hybrida* (as *Cineraria* ×*hybrida*) was one of the most popular early spring-flowering indoor plants of the 19th century; a hybrid between two species from the Canary Islands, it has recently had something of a revival. *Calceolaria* is the 300-species South American genus behind the *Calceolaria* Herbeohybrida Group, which was of great importance in the 19th century; the pocketbook plant is still around but in no great quantity.

Among hardy perennials, two Asteraceae genera are worth mention: *Gaillardia* and *Tanacetum*. Both are short-lived and require constant repropagation to survive in cultivation. Judging by the amount of coverage they had in the gardening press and the space devoted to them in nursery catalogues, they were among the top five perennials in early-20th-century Britain. Now, very few homeowners have even heard of them: a handful of *Tanacetum* cultivars are available, and *Gaillardia* exists as a number of seed strains. I am tempted to say that in a way it is not surprising they have all but vanished, as they lack the advantages of either clonal perennials or annuals; however, I note the growing importance of another daisy genus, *Echinacea*, which is chock full of heavily promoted cultivars with short lifespans, so maybe there is a place for such plants after all.

Tanacetum varieties in the Kelway
nursery's fields in Somerset, England,
1924. Courtesy of Kelways Plants Ltd.

Further Reading and Sources

I hope this book has encouraged every reader to further explore garden plant origins. The following is not intended as an exhaustive list; it is simply the resources I found most useful (and quote from directly) and which I would recommend to others who wish to delve deeper to turn to first of all. In particular, many of the books have extensive bibliographies or lists of cited sources, so readers are able to follow up on what interests them. For tracking down books on certain genera, and local libraries that hold them, *The OCLC Worldcat* is a most useful global library catalogue, online at worldcat.org.

For those who want botanical definitions of genera, Phillips and Rix (2002) have provided an invaluable guide. Finding out about the evolutionary origins of the species from which garden plants have been developed is fascinating but challenging—this is the field of phylogeny. The major source is the *Angiosperm Phylogeny Website*, mobot.org/mobot/research/apweb/welcome.html; this is a monumental undertaking by the Missouri Botanical Garden. Be warned—it is very heavy going, with an enormous amount of new botanical and other jargon to learn, but potentially very rewarding to those who persevere.

A very general source that has proved useful for "where does it grow?" type information is the compilation of material made by Andrew Chatto, the late husband of British plantswoman Beth Chatto, culled from a wide range of journals. Some years ago I helped Beth to make it available, so it is now online at bethchatto.co.uk/andrew-chatto-s-archives.htm. For those who want to find out more about the competitor/stress-tolerant model of plant survival strategies, which I and many others have found immensely useful in understanding plants, Grime (2001) offers the key text.

As I have said previously, I have consciously steered clear of getting too involved in the murky world of herbal and traditional medicine. There is a vast amount of material online, much of it very unreliable or downright dubious. Health claims can be relatively easily checked through a very useful and accessible medical website, *The Cochrane Collaboration*, online at cochrane.org. For a more general overview of plant uses, for food, materials, and medicine, *Plants for a Purpose*, online at pfaf.org, is extremely useful and very extensive in its referencing, although very much European and North American in focus; it compiles data from a wide range of sources, and in many ways is a model of what the Internet can do for researchers.

For the plant history material, two books, written half a century ago, cover most hardy garden plants, with erudition and occasional biting wit: the two titles by Coats. Even more information-packed, indeed a true work of scholarship, is the volume, in German, by Krausch. I have been able to include a great deal here on the enormously influential Japanese input into our range of garden plants because several years ago I discovered a remarkable and beautiful book, which I draw on extensively: Kashioka and Ogisu (1997).

The quarterly journal of the (UK) Hardy Plant Society is useful as a source covering the last 50 years or so, a period of rapid change in the world of perennial growing. The earlier volumes are all online at hardy-plant.org.uk/library/bulletin/index.php. I have found the world of German horticulture a particularly fascinating one to explore; there is very good website, *Garten-Literatur,* which acts as an entry portal for a vast array of sources, including a very good section on the key figure of Karl Foerster, online at garten-literatur.de.

Finally, I welcome readers to contact me via the Internet and send in further suggestions of reference sources. The links page on my website contains a section for additional reference material.

Angier, Bradford. 1974. *Field Guide to Edible Wild Plants.* Stackpole Books.

BBC Sussex. 2010. Chef cooks up recipes for Japanese knotweed. 21 April.

Beeton, Samual Orchart. 1874. *Beeton's Gardening Book.* Ward Lock.

Bell, M. 2000. *The Gardener's Guide to Growing Temperate Bamboos.* Timber Press.

Blankaart, Steven. 1698. *Dutch Herbal Plants and Spices.* Jan Claesz ten Hoorn.

Bowles, E. A. 1914. *My Garden in Summer.* T. C. and E. C. Jack.

Campbell-Culver, M. 2001. *The Origin of Plants.* Headline.

Coats, Alice M. 1956. *Flowers and Their Histories.* Hulton Press.

——. 1963. *Garden Shrubs and Their Histories.* Vista Press.

Cobbett, William. 1829. *The English Gardener.* Privately published.

Coffey, T. 1993, *The History and Folklore of North American Wildflowers.* Facts on File.

Compton, J. A. et al. 1998. Reclassification of *Actaea* to include *Cimicifuga* and *Souliea* (Ranunculaceae). *Taxon* 47(3):593–634.

Coombes, A. J. 2012. *The A to Z of Plant Names.* Timber Press.

Culpeper, Nicholas.1652. *The English Physician.* "Printed for the author."

Davies, Dilys. 1992. *Alliums: The Ornamental Onions.* Batsford.

Dirr, Michael. 1998. *Manual of Woody Landscape Plants.* Stipes Publications.

Duthweiler, S. 2011. *Neue Pflanzen für neue Gärten* (New plants for new gardens). Wernersche Verlagsgesellschaft.

Erichsen-Brown, C. 1989. *Medicinal and Other Uses of North American Plants.* Dover.

Evelyn, John. 1664. *Silva: or, a Discourse of Forest Trees.* John Martyn.

Farrer, Reginald. 1919. *The English Rock Garden.* 2 vols. T. C. and E. C. Jack.

Foerster, Karl. 1957. *Einzug der Gräser und Farne in die Gärten* (The advent of grasses and ferns in gardens). Neumann Verlag.

Genders, R. 1977. *Scented Flora of the World.* R. Hale.

Gerritsen, Henk, and Piet Oudolf. 2000. *Dream Plants for the Natural Garden.* Timber Press.

Gilpin, William. 1837. *Woodland Gleanings: Being an account of British forest-trees, Indigenous and Introduced.* Adam Scott.

Glenny, George. 1855. *Glenny's Hand-book of Practical Gardening.* C. Cox.

Gray, Jane Loring, ed. 1894. *Letters of Asa Gray.* Houghton, Mifflin, and Co.

Griesbach, R.J. 2002. Development of phalaenopsis orchids for the mass-market. Online at hort.purdue.edu/newcrop/ncnu02/v5-458.html.

Grime, J. P. 2001. *Plant Strategies, Vegetation Processes, and Ecosystem Properties.* Wiley.

Habermas, Jürgen. 2007. Obituary for American philosopher Richard Rorty. *Süddeutsche Zeitung,* 11 June.

Hibberd, Shirley. 1858. *The Floral World and Garden Guide.* Groombridge and Sons.

——. 1872. *The Ivy, a Monograph.* Groombridge and Sons.

——. [1879.] *Familiar Garden Flowers.* Cassell.

Hirtz, F. 2014. *Victor Lemoine*. Vent d'Est.

Hooker, William Jackson. 1840. *Flora Boreali-Americana*. Henry G. Bohn.

Hoot, S. B., et al. 1994. Phylogenetic relationships in *Anemone* (Ranunculaceae) based on morphology and chloroplast DNA. *Systematic Botany* 19(1):169–200.

Jacob, John. 1836. *West Devon and Cornwall Flora*. Longman, Rees, Orme, Brown, Green and Longman.

Jekyll, Gertrude. 1899. *Wood and Garden*. Longmans, Green, and Co.

Kamano, Johnny, Billy Faber, and Maurie Hartmann. 1946. I'm a lonely little petunia. Rytvoc Inc.

Kashioka, Seizo, and Mikinori Ogisu. 1997. *History and Principle of the Traditional Floriculture in Japan*. Kansai Tech Corporation.

Kerr, George W. 1910. *Sweet Peas Up-To-Date*. W. Atlee Burpee & Co.

Kilpatrick, J. 2007. *Gifts from the Gardens of China*. Frances Lincoln.

Kingsbury, N. 2009. *Hybrid: The History and Science of Plant Breeding*. University of Chicago Press.

Krausch, Heinz-Dieter. 2003. *Kaiserkron und Päonien rot: Entdeckung und Einführung unserer Gartenblumen* (Crown imperials and red peonies: the discovery and introduction of our garden flowers). Deutsche Taschenbuch Verlag.

Lloyd, Christopher. 1970. *The Well-Tempered Garden*. Collins.

Loudon, Jane. 1841. *The Ladies' Companion to the Flower Garden*. William Smith.

Maddock, James. 1822. *The Florist's Directory, a Treatise on the Culture of Flowers*. J. Green for John Harding.

Maxwell, Douglas Fyfe. 1927. *The Low Road: Hardy Heathers and the Heather Garden*. Sweet & Maxwell.

Novikoff, A. V., and J. Mitka. 2011. Taxonomy and ecology of the genus *Aconitum* L. in the Ukrainian Carpathians. *Wulfenia* 18:37–61.

Osler, Mirabel. 1997. *A Gentle Plea for Chaos*. Bloomsbury.

Pacific Bulb Society. pacificbulbsociety.org.

Pfeil, B.E., et al. 2002. Phylogeny of *Hibiscus* and the tribe Hibisceae (Malvaceae) using chloroplast DNA sequences of ndhF and the rpl16 intron. *Systematic Botany* 27(2):333–350.

Phillips, Roger, and Martyn Rix. 2002. *The Botanical Garden*. 2 vols. Macmillan.

Phillips, S. 2012. *An Encyclopaedia of Plants in Myth, Legend, Magic and Lore*. Robert Hale.

Plantzafrica (South African native plants). pza.sanbi.org.

Robinson, William. 1921. *The English Flower Garden*. John Murray.

Royal Horticultural Society. 1997. *The Garden* 122:420.

Sackville-West, Vita. 1951. *Your Garden*. Michael Joseph.

Shephard, S. 2003. *Seeds of Fortune*. Bloomsbury.

Silva-Tarouca, Ernst, et al. 1922. *Unsere Freiland-Stauden* (Our hardy perennials). Hölder-Pichler-Tempsky.

Slade, Naomi. 2014. *The Plant Lover's Guide to Snowdrops*. Timber Press.

Swindells, P. 1983. *Waterliles*. Croom Helm.

Switzer, Stephen. 1718. *Ichnographia Rustica*. For D. Brown et al.

Thomas, Graham Stuart. 1976. *Perennial Garden Plants*. Dent.

Weathers, John. 1904. *Beautiful Garden Flowers for Town and Country*. Simpkin, Marshall, Hamilton, Kent & Co.

——. 1905. *Beautiful Bulbous Plants for the Open Air*. Simpkin, Marshall & Co.

Woodcock, H. D., and W. T. Stearn. 1950. *Lilies of the World*. Country Life.

Photo Credits

akg-images, page 41.

Chris **Callard**, pages 282, 283, 284, and 285.

Lorenza **Figari**, pages 37, 88, and 110.

James **Hitchmough**, page 189.

Catherine **Lucas**, page 111.

Larry **Mellichamp**, pages 39, 55, 155, 317, and 328.

Paul **Picton**, pages 53 and 54.

Amalia **Robredo**, page 253.

Eric **Spruit**, page 136.

Bob **Wallis**, pages 45, 96, 129, 170, and 185.

All other photos are by the author.

Acknowledgments

This book was an idea that Anna Mumford, the former London commissioning editor for Timber Press, immediately "got." So thank you to her and to Andrew Beckman at Timber, along with Franni Farrell, Sarah Milhollin, and others at the press, for making it happen. I would like to thank my wife, Jo Eliot, for putting up with my continuous commentary on the progress of the book, particularly concerning the illustrations, the sourcing of which was a major learning curve. I would also like to thank Catherine Lucas, for reading through some key parts of the document; Yuko Tanabe Nagamura, for her assistance with translation of some of the Japanese material on chrysanthemums; and Diana Sessarego, our part-time gardener, for her help in the administration of the image selection. I would also like to thank the staff of the Royal Horticultural Society's Lindley Library for their help in my research and image sourcing.

To illustrate the book, I had to find material that I could use within the restrictions of copyright and which was affordable. Much valuable material is held by libraries who charge a license fee for reproduction, but who often have a poor understanding of the laws of supply and demand, and who therefore cost themselves out of the market. Often it proved cheaper to buy antique prints or old nursery catalogues and scan them myself—a task where the guidance of Sarah Milhollin at Timber has helped a great deal.

In contrast to the licensing system, a number of institutions are digitising their collections and making copies available for free, under public domain or creative commons agreements. The United States Library of Congress is one such—their online collection has proved very useful. One of the greatest pleasures of working on this book has been researching pictures from Amsterdam's Rijksmuseum. The museum has a wonderful online facility, which I think is one of the best things on the web: Rijkstudio, where you can build up your own collection of images of the museum's collection and divide them up into thematic sections. Images may be downloaded and on request, large file sizes, suitable for publication, will be provided by the museum, all at no cost. So, a big thank you to the museum's visionary policy and its staff.

I requested the assistance of a number of colleagues and others in providing me with images, either photographs of plants in the wild, or the right to use historic images, so I owe big thank-yous to those who responded, even if eventually I was not able to use all those generously provided: Lorenza Figari, Larry Mellichamp, Chris Callard, Bob Wallis, Bettina Jacobi, Constantin Bartning, Amalia Robredo, Ivor Stokes, Sophie Hughes, James Hitchmough, Robert Smaus, Tony Hayter, Jennifer Trehane, Paul Picton, Eric Spruit; to Kate West and Sarah Wilmot at the library of the John Innes Centre in Norfolk, England; to Ayako Nagase at Tokyo City University and the staff at Chiba University Library; to Andy Martin at Kelways Nursery. Finally, thanks to Paula Elliott, who acted as post restante for me in New York.

Index